Solution Manual

Partial Differential Equations for Scientists and Engineers

by
S. J. Farlow

Solution Manual

Partial Differential Equations for Scientists and Engineers
S. J. Farlow

Section 1: Introduction

Lesson 1: Introduction   4

Section 2: Diffusion Type Problems

Lesson 2: Diffusion Type Problems (Parabolic Equations)   10
Lesson 3: Boundary Conditions for Diffusion Type Problems   15
Lesson 4: Derivation of the Heat Equation   19
Lesson 5: Separation of Variables   24
Lesson 6: Transforming Nonhomogeneous BCs to Homogeneous Ones   30
Lesson 7: Solving More Complicated Problems by Separation of Variables   37
Lesson 8: Transforming Hard Problems into Easier Ones   43
Lesson 9:   Solving Nonhomogeneous PDEs (Eigenfunction Expansions)   51
Lesson 10: Integral Transforms (Sine and Cosine Transforms)   59
Lesson 11: The Fourier Series and Transform   71
Lesson 12: The Fourier Transform and Its Applications to PDEs   81
Lesson 13: The Laplace Transform   86
Lesson 14: Duhamel's Principle   91
Lesson 15: The Convection Term u_x in Diffusion Problems   97

Take a break and scurry on over to Page 102 and take in my amusing story, *Ten Reasons for Not Naming Your Cat Calculus*. The story is taken from my (moderately priced) book of the same name which is available at Amazon.com in paperback, kindle and Ipad formats. Enjoy.

Section 3: Hyperbolic Type Problems

Lesson 16: The One Dimensional Wave Equation (Hyperbolic Equations)   108
Lesson 17: The D'Alembert Solution of the Wave Equation   111
Lesson 18: More on the D'Alembert Equation   115
Lesson 19: Boundary Conditions Associated with the Wave Equation   121
Lesson 20: The Finite Vibrating String (Standing Waves)   125

Lesson 21: The Vibrating Beam (Fourth Order PDE)   134
Lesson 22: Dimensionless Problems   142
Lesson 23: Classification of PDEs (Canonical Form of Hyperbolic Equations)   149
Lesson 24: The Wave Equation in Two and Three Dimensions (Free Space)   155
Lesson 25: The Finite Fourier Transforms (Sine and Cosine Transforms)   151
Lesson 26: Superposition (The Backbone of Linear Systems)   167
Lesson 27: First Order Equations (Method of Characteristics)   171
Lesson 28: Nonlinear First-Order Equations (Conservation Equations)   178
Lesson 29: Systems of PDEs   184
Lesson 30: The Vibrating Drumhead (Wave Equation in Polar Coordinates   190

Stop and relax at this point in the book (Page 198) and take in my amusing story, Professor Pickle and the Omicron Affair. The story is taken from my (moderately priced) online book of the same name, which is available at Amazon.com in paperback, kindle and Ipad formats. Enjoy

Section 4: Elliptic Type Problems

Lesson 31: The Laplacian (an intuitive description)   205
Lesson 32: General Nature of Boundary Value Problems   215
Lesson 33: Interior Dirichlet Problem for a Circle   221
Lesson 34: The Dirichlet Problem in an Annulus   233
Lesson 35 Laplace's Equation in Spherical Coordinates (Spherical Harmonics)   239
Lesson 36: A Nonhomogeneous Dirichlet Problem (Green's Function)   248

Stop and relax at this point (Page 254) and take in my amusing story, *The God Equation: Sentry at the Pearly Gates*. The story is taken from my (moderately priced) online book: *Mathematics: Ain't There an App for That?*. The book is available at Amazon.com in paperback, kindle and Ipad formats. Enjoy

Section 5: Numerical and Approximate Methods

Lesson 37: Numerical Solutions (Elliptic Problems)   264
Lesson 38: An Explicit Finite-Difference Method   270
Lesson 39: An Implicit Finite-Difference Method (Crank-Nicolson Method)   273
Lesson 40: Analytic versus Numerical Solutions   276
Lesson 41: Classification of PDEs (Parabolic and Elliptic Equations)   280
Lesson 42: Monte Carlo Methods (An Introduction)   286
Lesson 43: Monte Carlo Solutions of PDEs   291
Lesson 44: Calculus of Variations (Euler-Lagrange Equations)   296
Lesson 45: Variation Methods for Solving PDEs (Method of Ritz)   301
Lesson 46: Perturbation Methods for Solving PDEs   307

Lesson 47: Conformal Mapping Solutions for PDEs

Stop and relax at this point in the book (Page 318) and take in my amusing story, *How I (Almost) Solved Fermat's Last Theorem*. The story is taken from my (moderately priced) online book: *The Girl Who Ate Equations for Breakfast*. which is available at Amazon.com in paperback, kindle and Ipad formats. Enjoy.

$$\Pi \Sigma \amalg \Sigma \Delta$$

Part 1: Introduction

Lesson 1: Introduction to Partial Differential Equations

> 1. Classify the following equations according to their order, linear or nonlinear, homogeneous or nonhomogenous, constant or variable coefficients, and if they are linear, whether they are elliptic, parabolic, or hyperbolic.
>
> $a)\ u_t = u_{xx} + 2u_x + u$
>
> $b)\ u_t = u_{xx} + e^{-t}$
>
> $c)\ u_{xx} + 3u_{xy} + u_{yy} = \sin x$
>
> $d)\ u_{tt} = uu_{xxxx} + e^{-t}$

Solution:

The general form of a second-order linear PDE in two independent variables is
$$Au_{xx} + Bu_{xy} + Cu_{yy} + Du_x + Eu_y + Fu = G$$

where A, B, C, D, E, F and G are functions of the independent variables, x, y in this case (which of course allows constant values as well). They are classified according to

$$\begin{cases} B^2 - 4AC < 0 \Rightarrow \text{ equation is elliptic} \\ B^2 - 4AC = 0 \Rightarrow \text{ equation is parabolic} \\ B^2 - 4AC > 0 \Rightarrow \text{ equation is hyperbolic} \end{cases}$$

a) $u_t = u_{xx} + 2u_x + u$ Second order, linear, homogeneous, constant coefficients, two independent variables x, t. Since the only nonzero second-order derivative is u_{xx}, we have $B = 0, C = 0$ and so $B^2 - 4AC = 0$ which

Lesson 1: Introduction to Partial Differential Equations

means the equation is parabolic. Parabolic equations describe diffusion phenomena, such as heat flow and diffusion of substances in a medium. This particular equation is called a one-dimensional heat (or diffusion) equation since it has 1 *space* dimension, in this case x.

b) $u_t = u_{xx} + e^{-t}$ Second order, linear, nonhomogeneous (the nonhomogeneous term is e^{-t}), constant coefficients, two independent variables x, t. Since the only second-order derivative is u_{xx}, we have $B = 0, C = 0$, hence $B^2 - 4AC = 0$, hence the equation is parabolic. Parabolic equations describe diffusion phenomena such as heat flow. This equation is called a one-dimensional heat (or diffusion) equation (dimension 1 since it only has 1 *space* dimension, in this case x). The difference of this equation and the one in a) is that this equation has an external heat source e^{-t} but does not have terms in u and u_x which describe other physical processes. More on these terms later.

c) $u_{xx} + 3u_{xy} + u_{yy} = \sin x$ Second order, linear, nonhomogeneous (the nonhomogeneous term is $\sin x$), constant coefficients, two independent variables x, y. Here $A = C = 1, B = 3$ and $B^2 - 4AC = 5 > 0$, hence the equation is hyperbolic for all x, y. PDEs with variable coefficients can be hyperbolic for some values of the independent variables x, y and parabolic or elliptic for other values of x, y.

d) $u_{tt} = uu_{xxxx} + e^{-t}$ Fourth order, nonlinear. The words homogenous or nonhomogeneous are not generally used in connection with nonlinear equations, nor are the words constant or non constant coefficients. The term that makes the equation nonlinear is uu_{xxxx}. Also, we don't (normally) classify PDEs as elliptic, parabolic, or hyperbolic unless they are linear. Almost all PDEs have either 1st or 2nd (partial) derivatives, but rarely higher than two, although the Koreweg-deVries equation $u_t + u_{xxx} - 6uu_x = 0$ is an important 3rd order (partial) derivative that describes soliton wave motion (google solitons).

> 2. How many solutions to the PDE $u_t = u_{xx}$ can you find? Try to find a solution of the form $u(x,t) = e^{ax+bt}$.

Solution: Of course, any linear homogeneous PDE always has the zero solution $u(x,t) \equiv 0$, but let's see if we can find nonzero solutions by trying an exponential solution of the form $u(x,t) = e^{ax+bt}$, where the idea is to plug this expression in the PDE and see if there exist numbers a,b that make the function satisfy the equation. Plugging this exponential in the PDE gives $be^{ax+bt} = a^2 e^{ax+bt}$ or $b = a^2$. In other words, any function of the form $u(x,t) = e^{ax+a^2 t} = e^{ax} e^{a^2 t}$ is a solution, where a is a real (or complex) number. (Plug it in the equation yourself to verify it is a solution.) For example $u = e^x e^t$, $u = e^{2x} e^{4t}$, $u = e^{-x} e^t$,... are all solutions. In fact if you let a be the complex constant $a = i$, then $b = i^2 = -1$ which means the complex function

$$u(x,t) = e^{ix} e^{-t}$$
$$= e^{-t}[\cos x + i \sin x]$$
$$= e^{-t} \cos x + i e^{-t} \sin x$$

is also a solution. In fact you can show (try it) that if a PDE has real coefficients (like the one in the problem) and if you have a complex solution, say $u = v(x,y) + iw(x,y)$, then both the real and complex parts v and w are also solutions. In this problem you can verify that both (real) functions $v(x,y) = e^{-t} \cos x$ and $w(x,y) = e^{-t} \sin x$ are solutions of the (one-dimensional heat) equation $u_t = u_{xx}$. Note that both these solutions describe decaying sine and cosine curves.

> 3. If $u_1(x,y)$ and $u_2(x,y)$ satisfy the homogeneous linear equation
>
> $$A u_{xx} + B u_{xy} + C u_{yy} + D u_x + E u_y + F u = 0$$
>
> where A, B, C, D, E, and F can either be constants or functions of x, y, then the sum $u_1 + u_2$ also satisfies the PDE.

Lesson 1: Introduction to Partial Differential Equations

Solution: Since u_1 and u_2 are solutions of the PDE, we have

$$A\left(\frac{\partial^2 u_1}{\partial x^2}\right) + B\left(\frac{\partial^2 u_1}{\partial x \partial y}\right)u_{xy} + C\left(\frac{\partial^2 u_1}{\partial y^2}\right)u_{yy} + D\left(\frac{\partial u_1}{\partial x}\right) + E\left(\frac{\partial u_1}{\partial y}\right) + Fu_1 = 0$$

$$A\left(\frac{\partial^2 u_2}{\partial x^2}\right) + B\left(\frac{\partial^2 u_2}{\partial x \partial y}\right)u_{xy} + C\left(\frac{\partial^2 u_2}{\partial y^2}\right)u_{yy} + D\left(\frac{\partial u_2}{\partial x}\right) + E\left(\frac{\partial u_2}{\partial y}\right) + Fu_2 = 0$$

Adding these equations, we find

$$A\left(\frac{\partial^2 u_1}{\partial x^2} + \frac{\partial^2 u_2}{\partial x^2}\right) + B\left(\frac{\partial^2 u_1}{\partial x \partial y} + \frac{\partial^2 u_2}{\partial x \partial y}\right)u_{xy} + C\left(\frac{\partial^2 u_1}{\partial y^2} + \frac{\partial^2 u_2}{\partial y^2}\right)u_{yy}$$

$$+ D\left(\frac{\partial u_1}{\partial x} + \frac{\partial u_2}{\partial x}\right) + E\left(\frac{\partial u_1}{\partial y} + \frac{\partial u_2}{\partial y}\right) + F(u_1 + u_2) = 0$$

and using the fact that the derivative of a sum is the sum of the derivatives, we have

$$A\left(\frac{\partial^2 (u_1 + u_2)}{\partial x^2}\right) + B\left(\frac{\partial^2 (u_1 + u_2)}{\partial x \partial y}\right)u_{xy} + C\left(\frac{\partial^2 (u_1 + u_2)}{\partial y^2}\right)u_{yy}$$

$$+ D\left(\frac{\partial (u_1 + u_2)}{\partial x}\right) + E\left(\frac{\partial (u_1 + u_2)}{\partial y}\right) + F(u_1 + u_2) = 0$$

which says that $u_1 + u_2$ is a solution of the PDE. Note that if the PDE was nonhomogeneous, i.e. $G \neq 0$, the sum $u_1 + u_2$ would not be a solution. You can also check and see that the sum of two solutions of a nonlinear PDE is not (always) a solution.

---

4. Probably the easiest of all PDEs to solve is

$$\frac{\partial u(x,y)}{\partial x} = 0$$

> which you have no doubt seen in a multi-variable calculus class. Can you find all the solutions of this equation?

Solution: Integrating this equation with respect to x yields $u(x,y) = g(y)$ where g is any (differentiable) function of y. You can see that $u(x,y) = \sin y$, $u(x,y) = e^y$, $u(x,y) = 10$ are typical solutions.

> 5. Find the solutions of the (second-order) cross-partial PDE
>
> $$\frac{\partial^2 u(x,y)}{\partial x \partial y} = \frac{\partial}{\partial x}\left(\frac{\partial u}{\partial y}\right) = 0$$

Solution: We first integrate the equation

$$\frac{\partial}{\partial x}\left(\frac{\partial u}{\partial y}\right) = 0$$

with respect to x, getting

$$\frac{\partial u}{\partial y} = g(y)$$

where $g(y)$ is an arbitrary (differentiable) function of y. We then integrate this equation with respect to y, which gives

$$u(x,y) = \int g(y)\,dy = \phi(y) + \psi(x)$$

where ϕ, ψ are arbitrary (differentiable) functions of y and x, respectively. In other words, the general solution consists of all sums of two (differentiable) functions ϕ, ψ where ψ depends only on x and ϕ depends only on y. For example $u = x + y, u = x^2 + \sin y,....$ are solutions of the PDE (check them), whereas $u(x,y) = xy$ is not.

Note: The above second-order PDE has two arbitrary *functions* in the solution, compared with the second-order ODE

$$\frac{d^2u(x)}{dx^2} = 0 \Rightarrow u(x) = c_1 x + c_2$$

that has two arbitrary *constants* c_1 and c_2. In other words, there are many more solutions of PDEs than of ODEs.

$$\Sigma\Pi\Omega\Delta\upsilon$$

Part 2 : Diffusion-Type Problems

Lesson 2 : Diffusion Type Problems (Parabolic Equations)

1. If the temperature $u(x,t)$ of an insulted rod is described by the PDE

$$u_t = \alpha^2 u_{xx}, \quad 0 \leq x \leq 1, \; 0 < t < \infty$$

and the initial temperature (IC) is

$$\text{IC}: u(x,0) = \sin(\pi x) \quad 0 \leq x \leq 1$$

and the BCs are

$$\text{BCs}: \begin{cases} u(0,t) = 0 \\ u(1,t) = 0 \end{cases} \quad 0 < t < \infty$$

What is the behavior of the rod temperature $u(x,t)$ at later values of time?

Solution : The rate u_t (degrees/unit time) which the temperature u increases or decreases is proportional to the concavity of the curve $u(x,t)$ as a function of x at every instant of time t. Hence the plots of the temperature for different values of time are shown below in Figure 2.1. We will see later that the function $u(x,t)$ that satisfies the PDE $u_t = \alpha^2 u_{xx}$ with given IC and BC is $u(x,t) = e^{-\alpha^2 t} \sin(\pi x)$, which evaluated at time $t = 0, 0.1, 0.2,$ and 0.3. For simplicity we pick the conductivity parameter $\alpha = 1$. Note that the temperature is gradually going to zero everywhere in the interval $0 \leq x \leq 1$. Not only is the temperature going to zero, but the 'profile' of the curve does not change, i.e. the profile $\sin(\pi x)$ is the function of x and it gets damped by the function of time $e^{-\alpha^2 t}$.

Lesson 2: Diffusion Type Problems (Parabolic Equations)

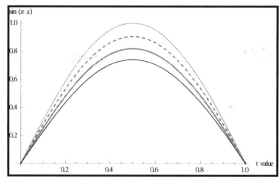

Figure 2.1: Temperature at times $t = 0, 0.1, 0.2, 0.3$

2. Suppose the rod in Problem 1 has a constant (internal) heat source everywhere along the rod so that the basic equation describing the heat flow within the rod is

$$u_t = \alpha^2 u_{xx} + 1, \; 0 < x < 1, \; 0 < t < \infty$$

Suppose we fix the boundary temperatures at each end $x=0$ and $x=1$ to be $u(0,t) = 0, \; u(1,t) = 1, \; 0 < t < \infty$. What will be the steady-state temperature of the rod (if there is one) after a long period of time? In other words, will the temperature of the rod approach a constant temperature $U(x)$ independent of time?

Solution : If there is a steady state solution, it will happen when the temperature, which we call $U(x)$, depends only depends on x and not time t. Hence, if there is a steady state temperature, we must have $u_t = 0$ which means the steady state temperature $U(x)$ should satisfy the ODE

$$\alpha^2 \frac{d^2 U(x)}{dx^2} + 1 = 0$$

which has the general solution

$$U(x) = -\frac{1}{2\alpha^2} x^2 + c_1 x + c_2$$

where c_1 and c_2 are arbitrary constants. If we determine the constants to fit the boundary conditions $U(0) = 0, \; U(1) = 1$, we get the steady state

12   Lesson 2: Diffusion Type Problems (Parabolic Equations)

$$U(x) = -\frac{1}{2\alpha^2}x^2 + \left(1 + \frac{1}{2\alpha^2}\right)x$$

When $\alpha = 1$ this steady state is shown in Figure 2.2.

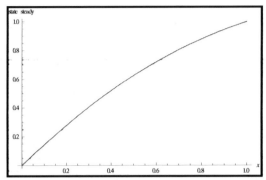

Figure 2.2: Steady state temperature of a rod

Starting with initial temperature of $u(x,0) = x$, the temperature at different values of time are shown in Figure 2.3.

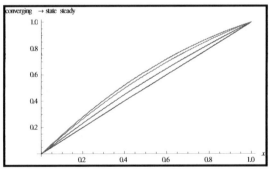

Figure 2.3 Solutions going from IC $u(x,0) = x$ to steady state

---

3. Suppose a metal rod loses heat across the lateral boundary according to the equation

$$u_t = \alpha^2 u_{xx} - \beta u, \quad 0 < x < 1, \quad \beta > 0$$

and suppose we keep the ends of the rod at $u(0,1) = 0$ and $u(1,t) = 1$. Find the steady-state temperature of the rod. Where is heat flowing in this problem?

Lesson 2: Diffusion Type Problems (Parabolic Equations)

Solution: Since the physical nature of the problem suggests a steady state, we set $u_t = 0$ the PDE reduces to the ODE

$$\alpha^2 \frac{d^2u}{dx^2} - \beta u = 0$$

and if we let the parameters $\alpha = \beta = 1$ for simplicity we have the solution

$$u(x) = c_1 e^x + c_2 e^{-x}$$

and plugging this into the BC $u(0) = 0, u(1) = 1$ one arrives at what we suspect is the steady state

$$u(x, \infty) = \left(\frac{e}{e^2 - 1}\right) e^x - \left(\frac{e}{e^2 - 1}\right) e^{-x}$$

We suspect that this is the steady state from the nature of the physical system. Heat flows into the right at the right hand side $(x=1)$ and flows through the rod but escapes on the lateral sides of the rod. However, keep in mind that at each point along the rod the net flow of heat in and out of the point is zero (i.e. net flow in = net flow out).

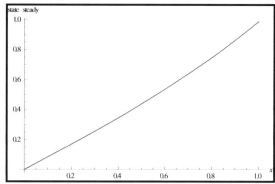

Figure 2.4 : Steady state of $u_t = \alpha^2 u_{xx} - \beta u$, $(\beta > 0)$ with BC
$u(0,t) = 0$, $u(1,t) = 1$

Although it is clear that the above function $u(x, \infty)$ satisfies the PDE as well as the boundary conditions, we must be careful using this technique of

Lesson 2: Diffusion Type Problems (Parabolic Equations)

setting $u_t = 0$ and solving the ODE since the solution may be unstable; i.e. nearby solutions move away from $u(x,\infty)$. Recalling from ODE the equation $y' - y = 1$, it is clear that $y(x) = -1$ is a constant solution, which we get from setting $y' = 0$, but $y(x) = -1$ is not the steady state of the equation since the general solution is $y(x) = ce^x - 1$; i.e. which means that all other solutions (for $c \neq 0$) move <u>away</u> from $y(x) = -1$ and so the equation has no steady state.

---

4. Suppose a laterally insulated rod of length $L=1$ has an initial temperature of $u(x,0) = \sin(3\pi x)$ and has its left and right ends kept at $u(0,t) = 0°C$ and $u(1,t) = 10°C$ respectively. What would be the IBVP that describes this problem?

Solution: The initial boundary value problem (IBVP) would be

$$\text{PDE: } u_t = \alpha^2 u_{xx} \quad 0 < x < 1, \ 0 < t < \infty$$

$$\text{BC: } \begin{cases} u(0,t) = 0, & 0 < t < \infty \\ u(1,t) = 10, & 0 < t < \infty \end{cases}$$

$$\text{IC: } \quad u(x,0) = \sin(3\pi x), \ 0 \leq x \leq 1$$

(Note that the PDE is defined inside the region $0 < x < 1$ and $0 < t < \infty$ since the derivatives u_t and u_{xx} are not defined on the boundaries.)

$$\Sigma\Omega\Upsilon\Delta\nabla$$

Lesson 3 : Boundary Conditions for Diffusion-Type Problems

> 1. Draw rough sketches of the solution of the following IBVP at different values of time? Will there be a steady-state temperature, and if so what is it?
>
> $$\text{PDE } u_t = \alpha^2 u_{xx}, \quad 0 < x < 200, \ 0 < t < \infty$$
>
> $$\text{BC } \begin{cases} u_x(0,t) = 0 \\ u_x(200,t) = -\dfrac{h}{k}\left[u(200,t) - 20\right] \end{cases} \quad 0 < t < \infty$$
>
> $$\text{IC } u(x,0) = 0°C \quad 0 \le x \le 200$$

Solution : The IBVP describes a rod that is insulated at the left end. At the right end an external heat source of 20 degrees is applied, which tends to keep temperature at 20 degrees. With this in mind our intuition tells us that the temperature of the rod will eventually be a constant of 20 degrees everywhere in the rod, i.e. $u(x,\infty) = 20$, $0 \le x \le 200$. An engineer might say that the right end of the rod has a sensor so when the temperature is less than 20 degrees, heat is applied increasing the temperature, and when the temperature is greater than 20 degrees, a cooling device will lower the temperature, the net effect being that the temperature is held around 20 degrees.

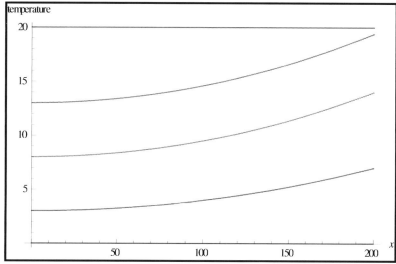

Figure 3.1: Solutions $u(x,0) = 0 \rightarrow u(x,\infty) = 20$

Note in the graphs in Figure 3.1 $u_x(0,t)=0$ for all time t and that $u_x(200,t)>0$ for all time t since $u(200,t)<20$.

---

2. What is your interpretation of the following IBVP:

$$\text{PDE } u_t = \alpha^2 u_{xx}, \quad 0<x<1,\ 0<t<\infty$$

$$\text{BC } \begin{cases} u(0,t)=0 \\ u_x(1,t)=1 \end{cases} \quad 0<t<\infty$$

$$\text{IC } u(x,0)=\sin(\pi x),\ 0\le x\le 1$$

Draw rough sketches of the solution. Does the solution have a steady state, and if so, what do you think it will be?

Solution: This IBVP might model a rod of length 1 which has an initial temperature in the shape of a sine curve. The temperature at the left end of the rod is fixed at zero, and heat flows *into* the right end of the rod (inward flux of 1 unit of heat per unit time - maybe cal/sec) at a constant rate. The first derivative u_x stands for *flux* in heat problems, if $u_x(x,t)>0$ heat is flowing to the *left* at point x and time t (from HOT to COLD), and if $u_x(x,t)<0$ heat is flowing to the *right* (from HOT to COLD). Intuitively, we suspect there will be a steady state, and if this is so then we must have $u_t=0$. Hence, the steady state would satisfy the boundary value problem

$$\frac{d^2 U}{dx^2}=0,\ U(0)=0,\ \frac{dU(1)}{dx}=1$$

which has the solution $U(x)=x$. You can draw the temperature curves starting at $u(x,0)=\sin(\pi x)$ and going to $u(x,\infty)=x$ in Figure 3.2.

Lesson 3: Boundary Conditions for Diffusion-Type Problems

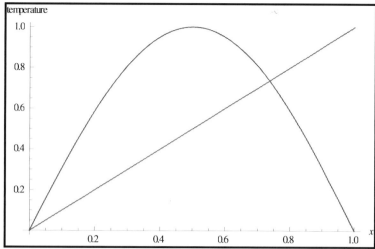

Figure 3.2 : Temperature going from $u(x,0) = \sin(\pi x)$ to $u(x,\infty) = x$

3. What is your physical interpretation of the following IBVP ?

PDE $u_t = \alpha^2 u_{xx}$, $0 < x < 1$, $0 < t < \infty$

BC $\begin{cases} u_x(0,t) = 0 \\ u_x(1,t) = 0 \end{cases}$ $0 < t < \infty$

IC $u(x,0) = \sin(\pi x)$, $0 \leq x \leq 1$

Make a rough sketch of the solution at various values of time? Does your intuition think there will be a steady-state solution, if so what is it?

Solution: This IBVP might be a model for an insulated rod of length 1, which has initial temperature $u(x,0) = \sin(\pi x)$ and is insulated at both the left and right ends. This being the case, there is no where for the heat energy in the rod (remember the rod is insulated on the outside) to go except to spread out, reaching a constant temperature, which should be the average value of the initial temperature (conservation of energy). That is

$$U(x) \equiv u(x,\infty) = \frac{1}{1} \int_0^1 \sin(\pi x) dx = \frac{2}{\pi}$$

Try drawing yourself various temperature curves for values of time between the initial temperature $u(x,0) = \sin(\pi x)$ and steady state temperature

18 Lesson 3: Boundary Conditions for Diffusion-Type Problems

$u(x,\infty) = 2/\pi$. (i.e. play Mother Nature yourself). Keep in mind that the solution must always have zero derivatives $u_x = 0$ at the end points $x = 0$ and $x = 1$.

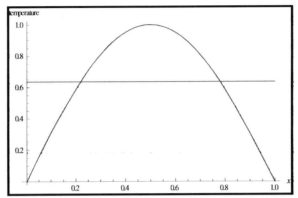

Figure 3.3 : Fill in temperatures $u(x,0) = \sin(\pi x) \rightarrow u(x,\infty) = 2/\pi$

---

4. Suppose a metal rod, laterally insulated, has an initial temperature of $20°C$, but immediately thereafter has one end fixed at $50°C$. The right end of the rod is immersed in a liquid solution of temperature $30°C$. What would be the IBVP that describes this problem?

Solution:

PDE $u_t = \alpha^2 u_{xx}$, $0 < x < 1$, $0 < t < \infty$

BC $\begin{cases} u(0,t) = 50 \\ u_x(1,t) = -\dfrac{k}{h}\left[u(1,t) - 30\right] \end{cases}$ $0 < t < \infty$

IC $u(x,0) = 20$, $0 \le x \le 1$

Note at the right end point $x = 1$ if the temperature inside (at the edge $x = 1$) is greater than 30 degrees, then $u_x(1,t) < 0$ which (according to Fourier's law) heat flow out of the region proportional to h/k.

$$\Sigma\Pi\Omega\Delta\amalg$$

Lesson 4 : Derivation of the Heat Equation

> 1. Substitute the units of each quantity $u, u_t, \ldots$ into the equation
>
> $$u_t = \alpha^2 u_{xx} - \beta u$$
>
> to see that every term has the units of $°C/\sec$.

Solution: The term u_t has units of deg/sec, and u_{xx} has units of deg/cm^2, and u has units of deg, so α^2 must have units of

$$\text{units}\left[\alpha^2\right] = \frac{\text{deg/sec}}{\text{deg/cm}^2} = \frac{\text{cm}^2}{\sec}$$

and β (time constant) has units of

$$\text{units}[\beta] = \frac{\text{deg/sec}}{\text{deg}} = \frac{1}{\sec}$$

> 2. Substitute the units of each quantity $u, u_t, \ldots$ into the equation
>
> $$u_t = \alpha^2 u_{xx} - v u_x$$
>
> where v has units of velocity to see that every term has the same units.

Solution : The term u_t clearly has units of deg/sec, u_{xx} has units of deg/cm^2, u has units of deg, and α^2 must have units of

$$\text{units}\left[\alpha^2\right] = \frac{\text{deg/sec}}{\text{deg/cm}^2} = \frac{\text{cm}^2}{\sec}$$

and v has the correct units of velocity

$$\text{units}[\nu] = \frac{\text{deg/sec}}{\text{deg/cm}} = \frac{\text{cm}}{\text{sec}}$$

> 3. Derive the heat equation
>
> $$u_t = \frac{1}{c\rho}\frac{\partial}{\partial x}\big[\kappa(x)u_x\big] + f(x,t)$$
>
> for the situation where the thermal conductivity $\kappa(x)$ depends on x.

Solution : The form of this PDE may look a little unusual to the reader, but this is a common way to write the partial derivative with respect to x when the proportionality constant $k = k(x)$ is a variable function of x. If k were a constant, then the equation would be

$$u_t = \frac{k}{c\rho}u_{xx} + f(x,t)$$

which may seem more common to the reader. To prove the one-dimensional heat equation when the thermal conductivity $k(x)$ varies, we begin with the conservation of energy relation (Eq 4.2 in the text on page 29)

$$c\rho A\int_x^{x+\Delta x} u_t(s,t)\,ds = A\big[k(x+\Delta x)u_x(x+\Delta x,t) - k(x)u_x(x,t)\big] + Ac\rho\int_x^{x+\Delta x} f(s,t)\,ds$$

which we now modify since the conductivity $k(x)$ is no longer a constant. This above equation is the mathematical way of saying that the net change of energy in $[x, x+\Delta x]$ is equal to the heat gain/loss across the boundaries at x and $x+\Delta x$ plus the heat generated in the interval $[x, x+\Delta x]$ by the heat source $f(x,t)$. We now do a little algebra and rewrite the above equation as

$$k(x+\Delta x)u_x(x+\Delta x,t) - k(x)u_x(x,t) = c\rho\int_x^{x+\Delta x}\big[u_t(s,t) - f(s,t)\big]\,ds$$

and apply the mean value theorem

Lesson 4 : Derivation of the Heat Equation

$$\int_a^b f(x)dx = f(\xi)(b-a), \; a < \xi < b$$

to the integral of this equation, getting

$$k(x+\Delta x)u_x(x+\Delta x, t) - k(x)u_x(x,t) = c\rho[u_t(\xi,t) - f(\xi,t)]\Delta x$$

We now solve for $u_t(\xi, t)$ getting

$$u_t(\xi,t) = \frac{1}{c\rho}\left\{\frac{k(x+\Delta x)u_x(x+\Delta x, t) - k(x)u_x(x,t)}{\Delta x}\right\} + f(\xi,t)$$

If we now let $\Delta x \to 0$ (which implies $\xi \to x$), we get the desired result

$$u_t(x,t) = \frac{1}{c\rho}\frac{\partial}{\partial x}[\kappa(x)u_x] + f(x,t)$$

Note: We could (although it is seldom written this way) write this last equation as

$$u_t(x,t) = \frac{1}{c\rho}\left[\frac{dk}{dx}\frac{\partial u}{\partial x} + k(x)\frac{\partial^2 u}{\partial x^2}\right] + f(x,t)$$

---

4. Suppose $u(x,t)$ measures the concentration of the substance in a moving stream moving (to the right) with velocity v, where now the concentration u changes both by convection and diffusion. Derive the equation

$$u_t = \alpha^2 u_{xx} - v u_x$$

from the fact that at any instant of time, the total mass of the material is not created or destroyed in the region $[x, x+\Delta x]$. Hint: Write the conservation equation

Lesson 4 : Derivation of the Heat Equation

> Change of mass inside $[x, x+\Delta x]$
> = Diffusion across boundaries
> + Material being carried the boundaries

Solution: We now must modify the conservation of energy equation, which says

Change of mass inside $[x, x+\Delta x]$
= Diffusion across boundaries
+ Material being carried the boundaries

to account for the fact that heat is lost (or gained) across the boundaries as a result of convection, not just diffusion as it was before. Hence, we have the new equation

$$c\rho A \int_{x}^{x+\Delta x} u_t(s,t)\,ds = Ak\left[u_x(x+\Delta x,t) - u_x(x,t)\right] - c\rho A v\left[u(x+\Delta x,t) - u(x,t)\right]$$

Notice the units of

$$c\rho A \int_{x}^{x+\Delta x} u_t(s,t)\,ds$$

are

$$\underbrace{\left(\frac{\text{cal}}{\text{deg}\cdot\text{gm}}\right)}_{c}\underbrace{\left(\frac{\text{gm}}{\text{cm}^3}\right)}_{\rho}\underbrace{(\text{cm}^2)}_{A}\underbrace{\left(\frac{\text{deg}}{\text{sec}}\right)}_{u_t}\underbrace{(\text{cm})}_{dx} = \frac{\text{cal}}{\text{sec}}$$

which has the same (cal/sec) units as

$$c\rho A v\left[u(x+\Delta x,t) - u(x,t)\right]$$

since

$$\underbrace{\left(\frac{\text{cal}}{\text{deg}\cdot\text{gm}}\right)}_{c}\underbrace{\left(\frac{\text{gm}}{\text{cm}^3}\right)}_{\rho}\underbrace{(\text{cm}^2)}_{A}\underbrace{\left(\frac{\text{cm}}{\text{sec}}\right)}_{v}\underbrace{(\text{deg})}_{u} = \frac{\text{cal}}{\text{sec}}$$

Lesson 4 : Derivation of the Heat Equation

and measures the heat gain or loss in $[x, x+\Delta x]$ as a result of convection. We now rewrite this equation as

$$k\left[u_x(x+\Delta x,t)-u_x(x,t)\right] = -c\rho v\left[u(x+\Delta x,t)-u(x,t)\right] + c\rho \int_x^{x+\Delta x} u_t(s,t)\,ds$$

and applying the mean value theorem to the integral, we have

$$k\left[u_x(x+\Delta x,t)-u_x(x,t)\right] = c\rho v\left[u(x+\Delta x,t)-u(x,t)\right] + c\rho u_t(\xi,t)\Delta x$$

Finally, solving for $u_t(\xi,t)$, we get

$$u_t(\xi,t) = \frac{k}{c\rho}\left\{\frac{u_x(x+\Delta x,t)-u_x(x,t)}{\Delta x}\right\} - v\left\{\frac{u(x+\Delta x,t)-u(x,t)}{\Delta x}\right\}$$

and letting $\Delta x \to 0$ which forces $\xi \to x$, we get the desired result

$$u_t = \alpha^2 u_{xx} - v u_x$$

where $\alpha^2 = k/c\rho$.

$$\Sigma\Pi\nabla\Omega\Sigma$$

Lesson 5: Separation of Variables

> 1. Show that
> $$u(x,t) = e^{-\lambda^2\alpha^2 t}\left[A\sin(\lambda x) + B\cos(\lambda x)\right]$$
> satisfies the PDE $u_t = \alpha^2 u_{xx}$ for arbitrary A, B.

Solution: Simply compute the partial derivatives

$$u_t = \left(-\lambda^2\alpha^2\right)e^{-\lambda^2\alpha^2 t}\left[A\sin(\lambda x) + B\cos(\lambda x)\right]$$

$$u_{xx} = e^{-\lambda^2\alpha^2 t}\left[-A\lambda^2\sin(\lambda x) - B\lambda^2\cos(\lambda x)\right]$$
$$= -\lambda^2 e^{-\lambda^2\alpha^2 t}\left[A\sin(\lambda x) + B\cos(\lambda x)\right]$$

$$\alpha^2 u_{xx} = \left(-\lambda^2\alpha^2\right)e^{-\lambda^2\alpha^2 t}\left[A\sin(\lambda x) + B\cos(\lambda x)\right]$$

Hence, we have $u_t = \alpha^2 u_{xx}$.

> 2. Prove the orthogonality property
> $$\int_0^1 \sin(m\pi x)\sin(n\pi x)\,dx = \begin{cases} 0 & m \neq n \\ 1/2 & m = n \end{cases}$$
> for the sine function. Hint: Use the identity
> $$\sin(m\pi x)\sin(n\pi x) = \frac{1}{2}\left[\cos(m-n)\pi x - \cos(m+n)\pi x\right]$$

Solution: We don't want to insult your intelligence by doing this trivial integral, but we will nevertheless.

Lesson 5: Separation of Variables   25

$(m=n)$ $\int_0^1 \sin(m\pi x)\sin(n\pi x)\,dx = \int_0^1 \sin^2(m\pi x)\,dx$

$$= \frac{1}{2}\int_0^1 \left[\cos(0)\pi x - \cos(2m\pi)x\right]dx$$

$$= \frac{1}{2}\int_0^1 \left[1 - \cos(2m\pi)x\right]dx$$

$$= \frac{1}{2}\left[x - \frac{1}{2m\pi}\sin(2m\pi)x\Big|_0^1\right]$$

$$= \frac{1}{2}$$

$(m \ne n)$ $\int_0^1 \sin(m\pi x)\sin(n\pi x)\,dx = \frac{1}{2}\int_0^1 \left[\cos(m-n)\pi x - \cos(m+n)\pi x\right]dx$

$$= \frac{1}{2}\left[\left(\frac{1}{(m-n)\pi}\right)\sin(m-n)\pi x\Big|_0^1\right]$$

$$-\frac{1}{2}\left[\left(\frac{1}{(m+n)\pi}\right)\sin(m+n)\pi x\Big|_0^1\right]$$

$$= 0$$

The last integral is zero since $\sin(k\pi x) = 0$ for any integer $k = 0, \pm 1, \pm 2, \ldots$.

---

3. Find the Fourier sine expansion of $\phi(x) = 1$, $0 \le x \le 1$. Draw the first three or four terms.

Solution : The Fourier (sine) series is

$$\phi(x) \equiv 1 = \sum_{n=1}^{\infty} a_n \sin(n\pi x)$$

To find the coefficients a_n, we multiply each side of the equation by $\sin(m\pi x)$ and integrate each side with respect to x from 0 to 1. Then, using the orthogonality following property of the sines and cosine functions

$$\int_0^1 \sin(m\pi x)\sin(n\pi x)\,dx = \begin{cases} 0 & m \neq n \\ 1/2 & m = n \end{cases}$$

$$\int_0^1 \cos(m\pi x)\cos(n\pi x)\,dx = \begin{cases} 0 & m \neq n \\ 1/2 & m = n \end{cases}$$

$$\int_0^1 \sin(m\pi x)\cos(n\pi x)\,dx = 0 \text{ all } m,n = 0,1,2,....$$

we get

$$\int_0^1 \sin(m\pi x)\,dx = a_m \int_0^1 \sin^2(m\pi x)\,dx$$

which we can solve for a_m, getting

$$a_m = \frac{\int_0^1 \sin(m\pi x)\,dx}{\int_0^1 \sin^2(m\pi x)\,dx} = \begin{cases} 0 & m = 0,2,4,... \\ \dfrac{4}{m\pi} & m = 1,3,5,... \end{cases}$$

Hence, the sine expansion of $f(x) = 1$, $0 \leq x \leq 1$ is

$$1 = \frac{4}{\pi}\left[\sin(\pi x) + \frac{1}{3}\sin(3\pi x) + \frac{1}{5}\sin(5\pi x) + \cdots\right],\ 0 \leq x \leq 1$$

Figure 5.1 Plotting the first four terms of this series we have

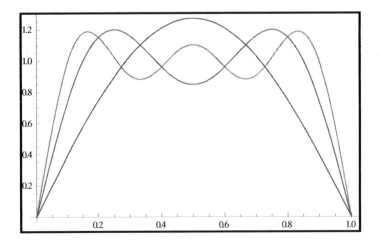

Figure 5.1 shows graphs of the 1st, 2nd, and 3nd terms of the Fourier series

Lesson 5: Separation of Variables

Note, if we were to find the Fourier *cosine* series

$$1 = \sum_{n=0}^{\infty} a_n \cos(n\pi x) = a_0 + a_1 \cos(\pi x) + a_2 \cos(2\pi x) + \cdots$$

of the function $f(x) = 1$, $0 \le x \le 1$, it would consist of one term, namely $a_0 = 1$. You should realize that since the Fourier series is periodic (period 2 in this case) that if we graphed the series on the entire real axis it would look like a square wave shown below in Figure 5.2.

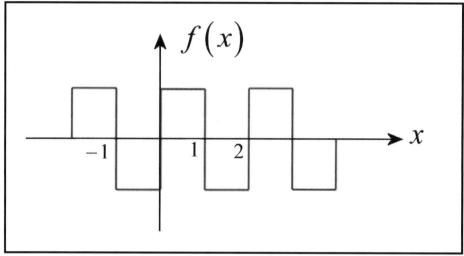

Figure 5.2 Fourier series a square wave outside the interval $[0,1]$

---

4. Using the results of Problem 3, what is the solution of the IBVP

$$\text{PDE } u_t = \alpha^2 u_{xx}, \quad 0 < x < 1, \ 0 < t < \infty$$

$$\text{BC } \begin{cases} u(0,t) = 0 \\ u(1,t) = 0 \end{cases} \quad 0 < t < \infty$$

$$\text{IC } u(x,0) = 1, \quad 0 \le x \le 1$$

Solution : We saw in Problem 3 that the IC $u(x,0) = 1$ has a Fourier sine series of

$$1 = \frac{4}{\pi}\left[\sin(\pi x) + \frac{1}{3}\sin(3\pi x) + \frac{1}{5}\sin(5\pi x) + \cdots\right], \quad 0 \le x \le 1$$

hence, the solution of the IBVP (choosing $\alpha = 1$) is

$$u(x,t) = \frac{4}{\pi}\left[e^{-\pi^2 t}\sin(\pi x) + \frac{1}{3}e^{-(3\pi)^2 t}\sin(3\pi x) + \frac{1}{5}e^{-(5\pi)^2 t}\sin(5\pi x) + \cdots\right]$$

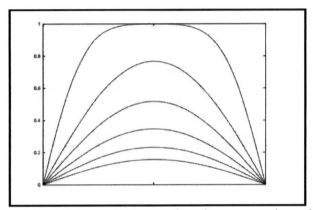

Figure 5.3: Temperature $u(x,0) \equiv 1 \to u(x,\infty) \equiv 0$

---

5. What is the solution to the IBVP in Problem 4 if the IC is changed to

$$u(x,0) = \sin(2\pi x) + \frac{1}{3}\sin(4\pi x) + \frac{1}{5}\sin(6\pi x)$$

Solution : When the IC is *already* a sine series, as it is in this problem, the solution is trivial, we simply slip the appropriate the exponential factor in each term. In this case, we have

$$u(x,t) = e^{-(2\pi\alpha)^2 t}\sin(2\pi x) + \frac{1}{3}e^{-(4\pi\alpha)^2 t}\sin(4\pi x) + \frac{1}{5}e^{-(6\pi\alpha)^2 t}\sin(6\pi x)$$

In general, the idea is to write the IC *as* a sine series and then include the factors of time in each term.

Lesson 5: Separation of Variables

> **6.** What is the solution to the IBVP in Problem 4 if the IC is changed to
>
> $$u(x,0) = x - x^2, \quad 0 \le x \le 1$$

Solution : The first step is to expand the IC $u(x,0) = x - x^2$ as a Fourier sine series

$$x - x^2 = \sum_{n=1}^{\infty} a_n \sin(n\pi x)$$

and then multiplying each side of the equation by $\sin(m\pi x)$ and integrating to get

$$\int_0^1 (x - x^2)\sin(m\pi x)\,dx = a_m \int_0^1 \sin^2(m\pi x)\,dx$$

or

$$a_m = \frac{\int_0^1 (x - x^2)\sin(m\pi x)\,dx}{\int_0^1 \sin^2(m\pi x)\,dx} = \begin{cases} 0 & n = 0, 2, 4, \ldots \\ \dfrac{8}{m^3 \pi^3} & m = 1, 3, \ldots \end{cases}$$

Hence, the sine expansion of the IC is

$$x - x^2 = \frac{8}{\pi^3}\left[\sin(\pi x) + \frac{1}{27}\sin(3\pi x) + \frac{1}{125}\sin(5\pi x) + \cdots\right]$$

and so the solution is

$$u(x,t) = \frac{8}{\pi^3}\left[e^{-(\pi\alpha)^2 t}\sin(\pi x) + \frac{1}{27}e^{-(3\pi\alpha)^2 t}\sin(3\pi x) + \frac{1}{125}e^{-(5\pi\alpha)^2 t}\sin(5\pi x) + \cdots\right]$$

$$\Sigma\Pi\sigma\Omega\Delta$$

Lesson 6 : Transforming Nonhomogeneous BCs Into Homogeneous Ones

1. Solve the following IBVP by transforming it into a new IBVP with homogeneous BCs and then solving the transformed problem. Does the solution agree with your intuition?

$$\text{PDE } u_t = \alpha^2 u_{xx}, \quad 0 < x < 1, \ 0 < t < \infty$$

$$\text{BC } \begin{cases} u(0,t) = 1 \\ u_x(1,t) + u(1,t) = 1 \end{cases} \quad 0 < t < \infty$$

$$\text{IC } u(x,0) = \sin(\pi x) + 1, \quad 0 \leq x \leq 1$$

Solution : Since the rod is fixed at 1 at the left end, and heat is added or subtracted respectively at the right end depending on whether the (outward normal) derivative $u_x(1,t) = 1 - u(1,t)$ is positive, heat applied if $u_x(1,t) > 0$, and heat extracted if $u_x(1,t) < 0$. Hence, our intuition tells us that the rod should reach a steady state temperature $u(x,\infty) = 1$. Hence, we try

$$u(x,t) = 1 + U(x,t)$$

where $U(x,t)$ is the transient solution. Hence

$$u_t = U_t$$
$$u_{xx} = U_{xx}$$
$$u(1,t) = 1 + U(1,t)$$
$$u_x(1,t) = U_x(1,t)$$
$$u(x,0) = 1 + U(x,0)$$

Plugging these value into the given IBVP, gives

Lesson 6: Transforming Nonhomogeneous BCs

PDE $U_t = \alpha^2 U_{xx},$ $0 < x < 1, \ 0 < t < \infty$

BC $\begin{cases} U(0,t) = 0 \\ U_x(1,t) + U(1,t) = 0 \end{cases}$ $0 < t < \infty$

IC $U(x,0) = \sin(\pi x),$ $0 \le x \le 1$

We can solve this problem by trying $U(x,t) = X(x)T(t)$, getting the two ODEs

$$T' + \lambda^2 T = 0$$

and

$$\text{ODE: } X'' + \lambda^2 X = 0$$

$$\text{BCs: } \begin{cases} X(0) = 0 \\ X'(1) + X(1) = 0 \end{cases}$$

The ODE in x has the solutions

$$X(x) = c_1 \cos \lambda x + c_2 \sin \lambda x$$

and plugging this in the BCs gives

$$X(0) = 0 \Rightarrow c_1 = 0$$
$$X'(1) + X(1) = \lambda \cos \lambda + \sin \lambda = 0 \Rightarrow \tan \lambda = -\lambda$$

Finding the roots of

$$\tan \lambda = -\lambda$$

we get an infinite number of positive roots

$$\lambda_1 = 2.02, \ \lambda_2 = 4.9, \ \lambda_3 = 7.9,...$$

illustrated in Figure 6.1.

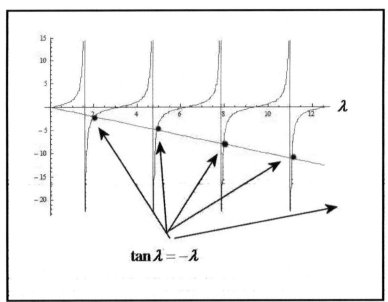

Figure 6.1 Graphical illustration of the roots of $\tan \lambda = -\lambda$

Hence we have the solutions

$$X(x) = \sin(\lambda_n x)$$
$$T_n(t) = e^{-\lambda_n t}$$

and thus

$$U(x,t) = \sum_{n=1}^{\infty} T_n(t) X_n(x) = \sum_{n=1}^{\infty} a_n e^{-\lambda_n t} \sin(\lambda_n x)$$

Note that we had to select only positive roots λ of $\tan \lambda = -\lambda$ otherwise the above infinite series for u would blow up. To find the coefficients a_n we plug this series in the IC, getting

$$U(x,0) = \sum_{n=1}^{\infty} a_n \sin(\lambda_n x) = \sin(\pi x) - x$$

We multiply each side of the series solution by $\sin(\lambda_m x)$ and integrate each side of the equation from 0 to 1. Doing this we get

$$\int_0^1 [\sin(\pi x) - x] \sin(\lambda_n x) dx = a_m \int_0^1 \sin^2(\lambda_m x) dx$$

from which we get

Lesson 6: Transforming Nonhomogeneous BCs   33

$$a_m = \frac{\int_0^1 [\sin(\pi x) - x] \sin(\lambda_m x) dx}{\int_0^1 \sin^2(\lambda_m x) dx}, \quad m = 1, 2, \ldots$$

whose values are $a_1 = 1.18, a_2 = 0.32, a_3 = 0.26, a_n = \cdots$ Hence

$$U(x,t) = \sum_{n=1}^{\infty} a_n e^{-\lambda_n t} \sin(\lambda_n x)$$
$$= 1.18 e^{-2.02 t} \sin(2.02 x) + 0.32 e^{-4.9 t} \sin(4.9 x) + \cdots$$

After we have found $U(x,t)$, we then compute

$$u(x,t) = 1 + U(x,t)$$
$$= 1 + \sum_{n=1}^{\infty} a_n e^{-\lambda_n t} \sin(\lambda_n x)$$
$$\doteq 1 + 1.18 e^{-2.02 t} \sin(2.02 x) + 0.32 e^{-4.9 t} \sin(4.9 x) + \cdots$$

---

2. Transform the following IBVP to an IBVP with zero BCs and solve the transformed problem. What does the solution look like for different values of time t. Does the solution agree with your intuition?

$$\text{PDE } u_t = u_{xx}, \quad 0 < x < 1, \ 0 < t < \infty$$

$$\text{BC } \begin{cases} u(0,t) = 0 \\ u(1,t) = 1 \end{cases} \quad 0 < t < \infty$$

$$\text{IC } u(x,0) = x^2, \quad 0 \leq x \leq 1$$

Solution : If we try a solution of the form $u(x,t) = x + U(x,t)$, we find $u_t = U_t(x,t)$, $u_x(x,t) = 1 + U_x(x,t)$, $u_{xx} = U_{xx}(x,t)$. Plugging these values into the IBVP, we obtain the new IBVP in $U(x,t)$:

$$\text{PDE } U_t = U_{xx}, \quad 0 < x < 1, \ 0 < t < \infty$$

$$\text{BC } \begin{cases} U(0,t) = 0 \\ U(1,t) = 0 \end{cases} \quad 0 < t < \infty$$

$$\text{IC } U(x,0) = x^2 - x, \quad 0 \leq x \leq 1$$

We now expand the IC $u(x,0) = x^2 - x$ as a Fourier sine series

$$x^2 - x = \sum_{n=1}^{\infty} a_n \sin(n\pi x)$$

and then multiplying each side of the equation by $\sin(m\pi x)$ and integrating to get

$$\int_0^1 (x^2 - x) \sin(m\pi x)\, dx = a_m \int_0^1 \sin^2(m\pi x)\, dx$$

or

$$a_m = \frac{\int_0^1 (x^2 - x)\sin(m\pi x)\, dx}{\int_0^1 \sin^2(m\pi x)\, dx} = \begin{cases} 0 & n = 0, 2, 4, \ldots \\ \dfrac{-8}{m^3 \pi^3} & m = 1, 3, \ldots \end{cases}$$

Hence, the sine expansion of the IC is

$$x^2 - x = -\frac{8}{\pi^3}\left[\sin(\pi x) + \frac{1}{27}\sin(3\pi x) + \frac{1}{125}\sin(5\pi x) + \cdots\right]$$

and so the solution is

$$u(x,t) = -\frac{8}{\pi^3}\left[e^{-\pi^2 t}\sin(\pi x) + \frac{1}{27}e^{-(3\pi)^2 t}\sin(3\pi x) + \frac{1}{125}e^{-(5\pi)^2 t}\sin(5\pi x) + \cdots\right]$$

Lesson 6: Transforming Nonhomogeneous BCs

3. Transform the following IBVP to one with zero BCs. Is the PDE in the transformed problem homogeneous?

$$\text{PDE } u_t = u_{xx}, \quad 0 < x < 1, \ 0 < t < \infty$$

$$\text{BC } \begin{cases} u_x(0,t) = 0 \\ u_x(1,t) + hu(1,t) = 1 \end{cases} \quad 0 < t < \infty$$

$$\text{IC } u(x,0) = \sin(\pi x), \ 0 \le x \le 1$$

Solution: Since the left end of the rod is insulated, and at the right end a heat source is applied (or subtracted) depending on if $1 - hu(1,t)$ is positive (heat applied since $u_x(1,t) > 0$) or negative (heat extracted) since $u_x(1,t) < 0$, our intuition tells us that the rod should reach a steady state temperature of the form $u(x,\infty) = Ax + 1$. Our intuition tells us that the temperature should approach some constant, say $u(x,\infty) = A$, thus we seek a solution as the sum of the steady state plus a transient. In other words

$$u(x,t) = A + U(x,t)$$

The first step is to find the steady state A which transforms the initial problem in u to a new problem in U with zero BC, thus making the problem easier to solve and finding the steady state in the process. Computing the derivatives of u, we have

$$u_t(x,t) = U_t(x,t)$$
$$u_x(x,t) = U_x(x,t)$$
$$u_{xx}(x,t) = U_{xx}(x,t)$$
$$u(x,0) = A + U(x,0)$$

and plugging them into the original IBVP, we find

PDE $U_t = U_{xx}$, $0 < x < 1$, $0 < t < \infty$

BC $\begin{cases} U_x(0,t) = 0 \\ U_x(1,t) + h[A + U(1,t)] = 1 \end{cases}$ $0 < t < \infty$

IC $A + U(x,0) = \sin(\pi x)$

If we now pick $A = 1/h$, we see that the BC at $x = 1$ becomes zero, and so the new IBVP becomes

PDE $U_t = U_{xx}$, $0 < x < 1$, $0 < t < \infty$

BC $\begin{cases} U_x(0,t) = 0 \\ U_x(1,t) + hU(1,t) = 0 \end{cases}$ $0 < t < \infty$

IC $U(x,0) = \sin(\pi x) - \dfrac{1}{h}$, $0 \leq x \leq 1$

Hence, we now solve this problem for $U(x,t)$ and then the solution of the original IBVP is

$$u(x,t) = \frac{1}{h} + U(x,t)$$

$\Sigma \Omega \mho \Sigma \Delta$

Lesson 7 : Solving More Complicated Problems by Separation of Variables

1. Solve the following IBVP.

$$\text{PDE } u_t = u_{xx}, \quad 0 < x < 1, \ 0 < t < \infty$$

$$\text{BC } \begin{cases} u(0,t) = 0 \\ u_x(1,t) = 0 \end{cases} \quad 0 < t < \infty$$

$$\text{IC } u(x,0) = x, \ 0 \le x \le 1$$

Solution : We have seen before that the Sturm-Liouville problem for this IBVP is

$$X'' + \lambda X = 0$$
$$X(0) = 0$$
$$X(1) = 0$$

which has eigenfunctions

$$X_n(x) = \sin\left(\frac{2n+1}{2}\right)\pi x, \ n = 0, 1, 2, \ldots$$

The first three ($n = 0, 1, 2$) are shown in Figure 7.1.

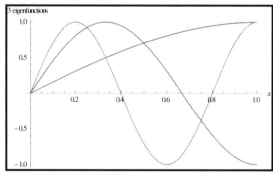

Figure 7.1: Eigenfunctions $\sin\left(\frac{\pi}{2}x\right)$, $\sin\left(\frac{3\pi}{2}x\right)$, $\sin\left(\frac{5\pi}{2}x\right)$

Hence, the solution of the IBVP is

Lesson 7: Solving More Complicated Problems

$$u(x,t) = \sum_{n=0}^{\infty} a_n e^{-\left(\frac{2n+1}{2}\right)^2 \pi^2 t} \sin\left(\frac{2n+1}{2}\right)\pi x$$

where the Fourier sine coefficients a_n satisfy the IC

$$x = \sum_{n=0}^{\infty} a_n \sin\left(\frac{2n+1}{2}\right)\pi x$$

To find the coefficients a_n, we multiply both sides of this equation by

$$\sin\left(\frac{2m+1}{2}\right)$$

and integrate both sides with respect to x from 0 to 1, getting

$$\int_0^1 x \sin\left(\frac{2m+1}{2}\right)\pi x\, dx = a_m \int_0^1 \sin^2\left(\frac{2m+1}{2}\right)\pi x\, dx$$

or

$$a_m = \frac{\int_0^1 x \sin\left(\frac{2m+1}{2}\right)\pi x\, dx}{\int_0^1 \sin^2\left(\frac{2m+1}{2}\right)\pi x\, dx} = \frac{(-1)^m 8}{(2m+1)^2 \pi^2}, \quad m = 0,1,2,...$$

(We know the functions

$$\left\{\sin\left(\frac{2m+1}{2}\right), m = 0,1,2,...\right\}$$

since they are the eigenvectors of a Sturm-Liouville BVP and the eigenvectors of a Sturm-Liouille system are always orthogonal). Hence, the solution is

Lesson 7 : Solving More Complicated Problems

$$u(x,t) = \sum_{n=0}^{\infty} a_n e^{-\left(\frac{2n+1}{2}\right)^2 \pi^2 t} \sin\left(\frac{2n+1}{2}\right)\pi x$$

$$\frac{8}{\pi^2}\left[e^{-\left(\frac{1}{2}\right)^2 \pi^2 t} \sin\left(\frac{\pi}{2}\right)x - e^{-\left(\frac{3}{2}\right)^2 \pi^2 t} \sin\left(\frac{3\pi}{2}\right)x + e^{-\left(\frac{5}{2}\right)^2 \pi^2 t} \sin\left(\frac{5\pi}{2}\right)x - \cdots \right]$$

---

2. What are the eigenvalues and eigenvectors of the following Sturm-Liouville problem.

$$\text{ODE: } X'' + \lambda X = 0, \ 0 < x < 1$$

$$\text{BC: } \begin{cases} X(0) = 0 \\ X'(0) = 0 \end{cases}$$

What are the functions $p(x)$, $q(x)$, and $r(x)$ in the general Sturm-Liouville problem?

Solution : The general Sturm-Liouville (eigenvalue/vector) problem has the form

$$\text{ODE:} \quad \frac{d}{dx}\left[p(x)\frac{dy}{dx}\right] - q(x)y + \lambda r(x) y = 0 \quad 0 < x < 1$$

$$\text{BCs:} \quad \begin{cases} \alpha_1 y(0) + \beta_1 y'(0) = 0 \\ \alpha_2 y(1) + \beta_2 y'(1) = 0 \end{cases}$$

In the current problem, we have $p(x) = 1$, $q(x) = 0$, $r(x) = 1$. For the BCs we have $\alpha_1 = \alpha_2 = 1$, $\beta_1 = \beta_2 = 0$. In the current problem, the general solution of the differential equation is

$$X(x) = c_1 \sin\sqrt{\lambda} x + c_2 \cos\sqrt{\lambda} x$$

and plugging this in the BC gives

Lesson 7: Solving More Complicated Problems

$$X(0) = c_2 = 0$$
$$X'(1) = c_1\sqrt{\lambda}\cos\sqrt{\lambda} - c_2\sqrt{\lambda}\sin\sqrt{\lambda} = 0$$

The second of the previous equations implies $\cos\sqrt{\lambda} = 0$, which gives us the eigenvalues of the problem

$$\sqrt{\lambda_n} = \left(\frac{2n+1}{2}\right)\pi, \; n = 0, 1, 2, \ldots$$

(we call $\sqrt{\lambda_n}$ the eigenvalues, not their square λ_n) and the eigenfunctions

$$X_n(x) = \sin\left(\frac{2n+1}{2}\right)\pi x, \; n = 1, 2, \ldots$$

The first 3 eigenfunctions are drawn in Figure 2.9 in Problem 1.

---

3. Solve the following problem with insulated boundaries. Does your solution agree with your interpretation of the problem? What is the steady-state solution and does it make sense?

$$\text{PDE } u_t = u_{xx}, \; 0 < x < 1, \; 0 < t < \infty$$

$$\text{BC } \begin{cases} u_x(0,t) = 0 \\ u_x(1,t) = 0 \end{cases} \; 0 < t < \infty$$

$$\text{IC } u(x,0) = x, \; 0 \leq x \leq 1$$

Solution: The Sturm-Liouville problem is

$$X'' + \lambda X = 0$$
$$X'(0) = 0$$
$$X'(1) = 0$$

which has eigenfunctions

Lesson 7 : Solving More Complicated Problems

$$X_n(x) = \cos(n\pi x), \; n = 0, 1, 2, \ldots$$

The first three ($n = 0, 1, 2$) are shown in Figure 7.2.

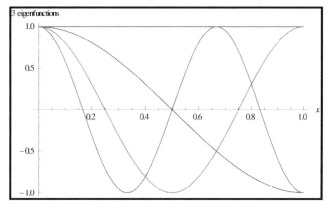

Figure 7.2: Eigenfunctions $1, \cos(\pi x), \cos(2\pi x), \cos(3\pi x), \ldots$

Hence, the solution of the IBVP is

$$u(x,t) = \sum_{n=0}^{\infty} a_n e^{-(n\pi)^2 t} \cos(n\pi x) = a_0 + a_1 e^{-(\pi)^2 t} \cos(\pi x) + \cdots$$

where the Fourier cosine coefficients a_n satisfy the IC

$$u(x,0) = x = \sum_{n=0}^{\infty} a_n \cos(n\pi x)$$

We find the Fourier coefficients a_n by multiplying both sides of this equation by $\cos(m\pi x)$ and integrating each side of the equation from 0 to 1, getting

$$\int_0^1 x \cos(m\pi x)\, dx = a_m \int_0^1 \cos^2(m\pi x)\, dx$$

or

$$a_m = \frac{\int_0^1 x \cos(m\pi x)\, dx}{\int_0^1 \cos^2(m\pi x)\, dx} = \begin{cases} 0 & m = 1, 3, \ldots \\ \dfrac{-4}{(m\pi)^2} & m = 0, 2, \ldots \end{cases}$$

Lesson 7: Solving More Complicated Problems

$$a_m = \frac{\int_0^1 x\cos(m\pi x)\,dx}{\int_0^1 \cos^2(m\pi x)\,dx} = \begin{cases} 1/2 & m=0 \\ 0 & m=1,3,\ldots \\ -4/(m\pi)^2 & m=2,4,\ldots \end{cases}$$

Hence, the solution

$$u(x,t) = \sum_{n=0}^{\infty} a_n e^{-(n\pi)^2 t}\cos(n\pi x)$$

$$= \frac{1}{2} - \frac{1}{\pi^2}\left[e^{-(2\pi)^2 t}\cos(2\pi x) + \frac{1}{4}e^{-(4\pi)^2 t}\cos(4\pi x) + \cdots\right]$$

Note as $t \to \infty$ the solution $u(x,t)$ converges to the constant

$$a_0 = \int_0^1 x\,dx = \frac{1}{2}$$

which you know from intuition since both ends of the rod are insulated and so the initial heat $u(x,0) = x$ must spread out along the rod at the average temperature, which is 1/2.

---

4. What are the eigenvalues and eigenvectors of the Sturm-Liouville problem?

ODE: $X'' + \lambda^2 X = 0,\ 0 < x < 1$

BC: $\begin{cases} X'(0) = 0 \\ X'(0) = 0 \end{cases}$

Solution: We have found the eigenvalues and eigenvectors in Problem 3 and found them to be

$$\text{eigenvalues}: \sqrt{\lambda_n} = n\pi,\ n = 0,1,2,\ldots$$
$$\text{eigenvectors}: X_n(x) = \cos(n\pi x),\ n = 0,1,2,\ldots$$

$$\Sigma\Omega\Upsilon\Pi\Delta$$

Lesson 8 : Transforming Harder Problems to Easier Ones

> 1. Solve the following IBVP.
>
> PDE $u_t = u_{xx} - u_x$, $\quad 0 < x < 1,\ 0 < t < \infty$
>
> BC $\begin{cases} u(0,t) = 0 \\ u(1,t) = 0 \end{cases} \quad 0 < t < \infty$
>
> IC $u(x,0) = e^{x/2}$, $\quad 0 \leq x \leq 1$
>
> by transforming it into an easier problem.

Solution : Since the IBVP describes the concentration $u(x,t)$ of a substance in a moving stream (due to the term $-u_x$) that is also diffusing (due to the term u_{xx}) in the medium, we might want to 'factor out' the amount of u that is due to the movement of the stream, and we do this by trying a solution of the form

$$u(x,t) = e^{(ax+bt)} U(x,t)$$

Computing the derivatives of u

$$u(x,t) = e^{(ax+bt)} U(x,t)$$
$$u_t(x,t) = b e^{(ax+bt)} U(x,t) + e^{(ax+bt)} U_t(x,t)$$
$$u_x(x,t) = a e^{(ax+bt)} U(x,t) + e^{(ax+bt)} U_x(x,t)$$
$$u_{xx}(x,t) = a^2 e^{(ax+bt)} U(x,t) + 2a e^{(ax+bt)} U_x(x,t) + e^{(ax+bt)} U_{xx}(x,t)$$

we plug these values into $u_t = u_{xx} - u_x$ getting

$$bU(x,t) + U_t(x,t) = a^2 U(x,t) + 2a U_x(x,t)$$
$$+ U_{xx}(x,t) - a U(x,t) - U_x(x,t)$$

or

Lesson 8: Transforming Harder Problems to Easier Ones

$$U_t = U_{xx} + (2a-1)U_x + (a^2 - a - b)U$$

We now set $2a - 1 = 0$, $a^2 - a - b = 0$ getting $a = \dfrac{1}{2}$, $b = -\dfrac{1}{4}$. Hence, the transformation

$$u(x,t) = e^{(0.5x - 0.25t)} U(x,t)$$

transforms the original PDE to $U_t = U_{xx}$ and the entire initial IBVP transforms to

$$\text{PDE } U_t = U_{xx}, \quad 0 < x < 1, \ 0 < t < \infty$$

$$\text{BC } \begin{cases} U(0,t) = 0 \\ U(1,t) = 0 \end{cases} \quad 0 < t < \infty$$

$$\text{IC } U(x,0) = 1, \quad 0 \le x \le 1$$

We solved this problem in Lesson 5 (Problem 4) and saw that the solution has the form

$$U(x,t) = \sum_{n=1}^{\infty} a_n e^{-(n\pi)^2 t} \sin(n\pi x)$$

where the coefficients a_n are determined so $U(x,t)$ satisfy the IC

$$U(x,0) = \sum_{n=1}^{\infty} a_n \sin(n\pi x) = 1$$

We find the coefficients a by multiplying each side of this equation by $\sin(m\pi x)$ and integrating x from 0 to 1, and using the orthogonality of the sine functions, getting

$$\int_0^1 \sin(m\pi x)\,dx = a_m \int_0^1 \sin^2(m\pi x)\,dx$$

Lesson 8: Transforming Harder Problems to Easier Ones

$$a_m = \frac{\int_0^1 \sin(m\pi x)\,dx}{\int_0^1 \sin^2(m\pi x)\,dx} = \begin{cases} 0 & m=0,2,4, \\ 4/m\pi & m=1,3,\ldots \end{cases}$$

Hence

$$U(x,t) = \sum_{n=1}^{\infty} a_n e^{-(n\pi)^2 t} \sin(n\pi x)$$

$$= \frac{4}{\pi} \sum_{n=1,3,\ldots}^{\infty} \frac{1}{n} e^{-(n\pi)^2 t} \sin(n\pi x)$$

$$= \frac{4}{\pi}\left[\sin(\pi x) + \frac{1}{3}\sin(3\pi x) + \frac{1}{5}\sin(5\pi x) + \cdots\right]$$

Hence, the solution $u(x,t)$ of the original IBVP is

$$u(x,t) = e^{\frac{1}{2}(x-t/2)} U(x,t)$$

$$= e^{\frac{1}{2}(x-t/2)} \sum_{n=1}^{\infty} a_n \sin(n\pi x)$$

$$= \frac{4}{\pi} e^{\frac{1}{2}(x-t/2)} \sum_{n=1}^{\infty} \frac{1}{(2n-1)} \sin[(2n-1)\pi x]$$

> **Note:** In general, the PDE $u_t = \alpha^2 u_{xx} - v u_x$ that contains a convection term $-v u_x$ (we include a negative with the term since that means the substance described by $u(x,t)$ moves in the direction of *increasing* x), can be transformed to the PDE $U_t = \alpha^2 U_{xx}$ by the transformation
>
> $$u(x,t) = e^{v\left(\frac{2x-vt}{4\alpha^2}\right)} U(x,t)$$
>
> The proof is identical to the steps carried out in this problem. That is, plug
>
> $$u(x,t) = e^{ax+bt} U(x,t)$$
>
> into $u_t = \alpha^2 u_{xx} - v u_x$ and find a,b so the new PDE in $U(x,t)$ does not contain the term involving U_x and U.

> Use steps a), b), and c) below to solve the IBVP
>
> $$\text{PDE } u_t = u_{xx} - u + x, \quad 0 < x < 1, \ 0 < t < \infty$$
>
> $$\text{BC } \begin{cases} u(0,t) = 0 \\ u(1,t) = 1 \end{cases} \quad 0 < t < \infty$$
>
> $$\text{IC } u(x,0) = 0, \ 0 \le x \le 1$$
>
> a) transforming the nonhomogeneous BCs to homogeneous ones
> b) transforming into a new PDE without the term $-u$.
> c) solving the resulting problem.

Solution: Since the temperature at the ends of the rod are fixed at 0 and 1, we suspect the temperature will reach a steady state of $u(x,\infty) = x$, Hence, we try a solution of the form

$$u(x,t) = x + U(x,t)$$

Plugging this value into the original IBVP, we arrive at (the x term in the PDE with cancel)

$$\text{PDE } U_t = U_{xx} - U \quad 0 < x < 1, \ 0 < t < \infty$$

$$\text{BC } \begin{cases} U(0,t) = 0 \\ U(1,t) = 0 \end{cases} \quad 0 < t < \infty$$

$$\text{IC } U(x,0) = -x, \ 0 \le x \le 1$$

Since the term $-U$ in this PDE is a damping (or evaporation) term, we try to find a solution of this equation of the form $U(x,t) = e^{-t} w(x,t)$. Plugging this into the previous IBVP we find

Lesson 8: Transforming Harder Problems to Easier Ones 47

$$\text{PDE } w_t = w_{xx} \quad 0 < x < 1, \ 0 < t < \infty$$

$$\text{BC } \begin{cases} w(0,t) = 0 \\ w(1,t) = 0 \end{cases} \quad 0 < t < \infty$$

$$\text{IC } w(x,0) = -x, \ 0 \le x \le 1$$

This IBVP has the solution

$$w(x,t) = \sum_{n=1}^{\infty} a_n e^{-(n\pi)^2 t} \sin(n\pi x) dx$$

where we find the coefficients a_n so the IC

$$w(x,0) = \sum_{n=1}^{\infty} a_n \sin(n\pi x) dx = -x$$

is satisfied. We do this by multiplying each side of the equation by $\sin(m\pi x)$ and integrating each side with respect to x from 0 to 1, getting

$$-\int_0^1 x \sin(m\pi x) dx = a_m \int_0^1 \sin^2(m\pi x) dx$$

$$a_m = -\frac{\int_0^1 x \sin(m\pi x) dx}{\int_0^1 \sin^2(m\pi x) dx} = (-1)^{m+1} \frac{2}{m\pi}, \quad m = 1, 2, 3, \ldots$$

Hence

$$w(x,t) = \sum_{n=1}^{\infty} a_n e^{-(n\pi)^2 t} \sin(n\pi x) dx$$

$$= \frac{2}{\pi} \sum_{n=1}^{\infty} \frac{1}{n} (-1)^{n+1} e^{-(n\pi)^2 t} \sin(n\pi x)$$

$$= \frac{2}{\pi} \left[e^{-(\pi)^2 t} \sin(\pi x) - \frac{1}{3} e^{-(3\pi)^2 t} \sin(3\pi x) + \frac{1}{5} e^{-(5\pi)^2 t} \sin(5\pi x) + \cdots \right]$$

and so

$$U(x,t) = e^{-t} w(x,t)$$

$$= e^{-t} \sum_{n=1}^{\infty} a_n e^{-(n\pi)^2 t} \sin(n\pi x) dx$$

$$= \frac{2}{\pi} e^{-t} \sum_{n=1}^{\infty} \frac{1}{n} (-1)^{n+1} e^{-(n\pi)^2 t} \sin(n\pi x)$$

$$= \frac{2}{\pi} e^{-t} \left[e^{-(\pi)^2 t} \sin(\pi x) - \frac{1}{3} e^{-(3\pi)^2 t} \sin(3\pi x) + \frac{1}{5} e^{-(5\pi)^2 t} \sin(5\pi x) + \cdots \right]$$

So, FINALLY, the solution $u(x,t)$ of the original IBVP is

$$u(x,t) = x + U(x,t)$$

$$= x + \frac{2}{\pi} e^{-t} \left[e^{-(\pi)^2 t} \sin(\pi x) - \frac{1}{3} e^{-(3\pi)^2 t} \sin(3\pi x) + \frac{1}{5} e^{-(5\pi)^2 t} \sin(5\pi x) + \cdots \right]$$

---

3. Solve

$$\text{PDE } u_t = u_{xx} - u \quad 0 < x < 1, \ 0 < t < \infty$$

$$\text{BC } \begin{cases} u(0,t) = 0 \\ u(1,t) = 0 \end{cases} \quad 0 < t < \infty$$

$$\text{IC } u(x,0) = \sin(\pi x), \ 0 \leq x \leq 1$$

directly by separation of variables without making any preliminary transformations. Does your solution agree with the solution you obtain if you make the transformation $u(x,t) = e^{-t} w(x,t)$ in advance?

Solution : Since the term $-u$ in this PDE is a damping term, we could solve this problem by seeking a solution of the form $u(x,t) = e^{-t} U(x,t)$. This factors out the damping so we are only left with diffusion described by the variable $U(x,t)$. Plugging this into the IBVP we find

Lesson 8: Transforming Harder Problems to Easier Ones   49

$$\text{PDE } U_t = U_{xx} \quad 0 < x < 1,\ 0 < t < \infty$$
$$\text{BC } \begin{cases} U(0,t) = 0 \\ U(1,t) = 0 \end{cases} \quad 0 < t < \infty$$
$$\text{IC } U(x,0) = \sin(\pi x),\ 0 \le x \le 1$$

which has the solution

$$U(x,t) = e^{-\pi^2 t} \sin(\pi x)$$

thus the original problem has the solution

$$\begin{aligned} u(x,t) &= e^{-t} U(x,t) \\ &= e^t e^{-\pi^2 t} \sin(\pi x) \\ &= e^{-(\pi^2 - 1)t} \sin(\pi x) \end{aligned}$$

However, if we solve this problem without the preliminary transformation, we use separation of variables, seeking a solution of the form $u(x,t) = X(x)T(t)$, and plugging this into the PDE we get

$$XT' = X''T - XT$$

and dividing by XT, arriving at

$$\frac{T'}{T} = \frac{X''}{X} - 1 = -\lambda^2$$

or

$$T' + \lambda^2 T = 0$$

$$X'' + (\lambda^2 - 1) X = 0$$
$$\begin{cases} X(0) = 0 \\ X'(1) = 0 \end{cases}$$

The solution of the ODE in $X(x)$ is

$$X(x) = c_1 \cos\left(\sqrt{\lambda^2 - 1}\, t\right) + c_2 \sin\left(\sqrt{\lambda^2 - 1}\, t\right)$$

and using the BCs give $c_1 = 0$ and

$$X(1) = c_2 \sin\left(\sqrt{\lambda^2 - 1}\right) = 0$$
$$\Rightarrow \sqrt{\lambda^2 - 1} = n\pi$$
$$\Rightarrow \lambda^2 = (n\pi)^2 + 1$$
$$\Rightarrow \lambda_n = \sqrt{(n\pi)^2 + 1}$$

Hence, we arrive at solutions

$$u_n(x,t) = X_n(x) T_n(t)$$
$$= e^{-\left[(n\pi)^2 + 1\right]t} \sin\left(\sqrt{(n\pi)^2 + 1}\, x\right)$$
$$= e^{-t}\left[e^{-(n\pi)^2 t} \sin(n\pi x)\right]$$

and so we get a solution of the form

$$u(x,t) = \sum_{n=1}^{\infty} a_n X_n(x) T_n(t)$$
$$= e^{-t} \sum_{n=1}^{\infty} a_n e^{-(n\pi)^2 t} \sin(n\pi x)$$

If we find the coefficients a_n so that $u(x,0) = \sin(\pi x)$ we find the only non zero value is $a_1 = 1$, and so

$$u(x,t) = e^{-t} \sin(\pi x)$$

which is the solution we got when we made the initial change of variables.

> **Note:** In general the PDE $u_t = \alpha^2 u_{xx} - \beta u$ with decay term βu can be transformed to the equation $U_t = \alpha^2 U_{xx}$ with the transformation $u(x,t) = e^{-\beta t} U(x,t)$. The decay constant β is generally positive since then a positive value of u means that $-\beta u < 0$ which makes u_t smaller.

$$\Sigma \Pi \Omega \Upsilon \Delta$$

Lesson 9: Solving Nonhomogeneous PDEs (Eigenfunction Expansions)

1. The solution of the IBVP

   PDE $u_t = u_{xx} + \sin(3\pi x)$, $0 < x < 1$, $0 < t < \infty$

   BC $\begin{cases} u(0,t) = 0 \\ u(1,t) = 0 \end{cases}$ $0 < t < \infty$

   IC $u(x,0) = \sin(\pi x)$, $0 \leq x \leq 1$

   is given in the book by

   $$u(x,t) = e^{-(\pi\alpha)^2 t}\sin(\pi x) + \frac{1}{(3\pi\alpha)^2}\left[1 - e^{-(3\pi\alpha)^2 t}\right]\sin(3\pi x)$$

   Does this solution agree with your intuition? What does the solution look like?

Solution: The solution starts off looking like $u(x,0) = \sin(\pi x)$ and gradually morphs into the steady state

$$u(x,\infty) = \frac{1}{(3\pi\alpha)^2}\sin(3\pi x)$$

2. Solve the following IBVP

   PDE $u_t = u_{xx} + \sin(\pi x) + \sin(2\pi x)$, $0 < x < 1$, $0 < t < \infty$

   BC $\begin{cases} u(0,t) = 0 \\ u(1,t) = 0 \end{cases}$ $0 < t < \infty$

   IC $u(x,0) = 0$, $0 \leq x \leq 1$

Solution:

Step 1: We expand the nonhomogeneous terms of the PDE

$$f(x,t) = \sum_{n=1}^{\infty} f_n(t)\sin(n\pi x) = \sin(\pi x) + \sin(2\pi x)$$

and so we identify

$$f_1(t) = f_2(t) \equiv 1$$
$$f_n(t) \equiv 0, \ n = 3,4,\ldots$$

Step 2: We now seek a solution of the form

$$u(x,t) = \sum_{n=1}^{\infty} T_n(t) X_n(x)$$

where (in this particular problem) the $X_n(x)$ are the eigenfunctions of the Sturm-Liouville problem

$$X'' + \lambda^2 X = 0$$
$$X(0) = 0$$
$$X(1) = 0$$

and the $T_n(t)$ are solutions of the initial-value problems

$$T_n' + (n\pi)^2 T_n = f_n(t)$$
$$T_n(0) = 0$$

Solving these problems, we find

$$X_n(x) = \sin(n\pi x), \ n = 1,2,\ldots$$
$$T_n(t) = \frac{1}{(n\pi)^2}\left[1 - e^{-(n\pi)^2 t}\right], \ n = 1,2.$$
$$T_n(t) = 0, \ n = 3,4,\ldots$$

Hence, the solution of the IBVP is

Lesson 9: Solving Nonhomogeneous PDES

$$u(x,t) = \sum_{n=1}^{\infty} T_n(t) X_n(x)$$

$$= \frac{1}{\pi^2}\left[1 - e^{-\pi^2 t}\right] \sin(\pi x) + \frac{1}{(2\pi)^2}\left[1 - e^{-(2\pi)^2 t}\right] \sin(2\pi x)$$

---

3. Solve the following IBVP by the method of eigenfunction expansion.

   PDE $u_t = u_{xx} + \sin(\pi x)$, $0 < x < 1$, $0 < t < \infty$

   BC $\begin{cases} u(0,t) = 0 \\ u(1,t) = 0 \end{cases}$ $0 < t < \infty$

   IC $u(x,0) = 1$, $0 \le x \le 1$

Solution: Step 1: We expand the nonhomogeneous terms of the PDE

$$f(x,t) = \sum_{n=1}^{\infty} f_n(t) \sin(n\pi x) = \sin(\pi x)$$

and so we identify

$$f_1(t) \equiv 1$$
$$f_n(t) \equiv 0, \ n = 2, 3, \dots$$

Step 2: We now seek a solution of the form

$$u(x,t) = \sum_{n=1}^{\infty} T_n(t) X_n(x)$$

where (in this particular problem) the $X_n(x)$ are the eigenfunctions of the Sturm-Liouville problem

$$X'' + \lambda^2 X = 0$$
$$X(0) = 0$$
$$X(1) = 0$$

and the $T_n(t)$ are solutions of the initial-value problems

$$T_n' + (n\pi)^2 T_n = 0$$
$$T_n(0) = a_n \quad n = 1, 2, 3, ...$$

where a_n are the Fourier coefficients in the Fourier series expansion of the IC $u(x,0) = 1$. That is

$$u(x,0) = 1 = \sum_{n=1}^{\infty} a_n \sin(x\pi x) \Rightarrow a_n = 2\int_0^1 \sin(n\pi x)\,dx = \frac{4}{(2n-1)\pi}$$

For this problem, we have

$$T_1' + \pi^2 T_1 = 1$$
$$T_1(0) = 2\int_0^1 \sin(\pi x)\,dx = \frac{4}{\pi}$$

$$T_n' + (n\pi)^2 T_n = 0$$
$$T_n(0) = 2\int_0^1 \sin(n\pi x)\,dx = \frac{4}{(2n-1)\pi} \quad n = 2, 3, ...$$

Solving these problems, we find

$$X_n(x) = \sin(n\pi x), \quad n = 1, 2, ...$$
$$T_1(t) = \frac{4}{\pi}e^{-\pi^2 t} + \frac{1}{\pi^2}\left[1 - e^{-\pi^2 t}\right].$$
$$T_n(t) = \begin{cases} 0, & n = 2, 4, 6, ... \\ \dfrac{4}{n\pi}e^{-(n\pi)^2 t} & n = 3, 5, 7,. \end{cases}$$

Hence, the solution of the IBVP is

$$u(x,t) = \sum_{n=1}^{\infty} T_n(t) X_n(x)$$

$$= \left\{ \frac{4}{\pi} e^{-\pi^2 t} + \frac{1}{\pi^2} \left[1 - e^{-\pi^2 t} \right] \right\} \sin(\pi x) + \frac{4}{\pi} \sum_{n=1}^{\infty} \frac{1}{(2n+1)} e^{-[(2n+1)\pi]^2 t} \sin\left[(2n+1)\pi x\right]$$

---

4. Solve the following IBVP by the method of eigenfunction expansion where λ_1 is the first root of the equation $\tan \lambda = -\lambda$. What are the eigenfunctions X_n of the problem?

$$\text{PDE } u_t = u_{xx} + \sin(\lambda_1 x), \quad 0 < x < 1, \; 0 < t < \infty$$

$$\text{BC } \begin{cases} u(0,t) = 0 \\ u_x(1,t) + u(1,t) = 0 \end{cases} \quad 0 < t < \infty$$

$$\text{IC } u(x,0) = 0, \; 0 \le x \le 1$$

Solution: Step 1: We expand the nonhomogeneous terms of the PDE

$$f(x,t) = \sum_{n=1}^{\infty} f_n(t) X_n(x)$$

where $X_n(x)$ are the eigenfunctions of

$$X'' + \lambda^2 X = 0$$
$$X(0) = 0$$
$$X'(1) + X(1) = 0$$

In this problem the nonhomogeneous term $\sin(\lambda_1 x)$ is the first eigenfunction and so we have $f_1(t) = 1, f_n(t) = 0, n = 2, 3, \ldots$ Hence, we seek a solution of the form

$$u(x,t) = \sum_{n=1}^{\infty} T_n(t) X_n(x)$$

where (in this particular problem) the $X_n(x)$ are the eigenfunctions of the Sturm-Liouville problem

$$X'' + \lambda^2 X = 0$$
$$X(0) = 0$$
$$X'(1) + X(1) = 0$$

which are

$$X_n(x) = \sin \lambda_n x$$

where λ_n is the nth root of $\tan \lambda = -\lambda$, and $T_n(t)$ are solutions of the initial-value problems

$$T_n' + \lambda_n^2 T_n = f_n(t) = \begin{cases} 1 & n=1 \\ 0 & n=2,3,... \end{cases}$$

$$T_n(0) = a_n$$

and a_n are the coefficients of the Fourier series expansion of the IC $u(x,0) = 0$. That is

$$u(x,0) = 0 = \sum_{n=1}^{\infty} a_n \sin(x\pi x) \Rightarrow a_n = 0$$

In this problem, we have

$$\begin{matrix} T_1' + \lambda_1^2 T_1 = 1 \\ T_1(0) = 0 \end{matrix} \Rightarrow T_1(t) = \frac{1}{\lambda_1}\left[1 - e^{-\lambda_1 t}\right]\sin(\lambda_1 x)$$

$$\begin{matrix} T_n' + \lambda_n^2 T_n = 0 \\ T_n(0) = 0 \end{matrix} \Rightarrow T_n(t) = 0 \quad n = 2,3,....$$

Hence, the solution of the IBVP is

$$u(x,t) = \sum_{n=1}^{\infty} T_n(t) X_n(x) = \frac{1}{\lambda_1}\left[1 - e^{-\lambda_1 t}\right]\sin(\lambda_1 x)$$

Lesson 9: Solving Nonhomogeneous PDES 57

> 5. Solve the IBVP
>
> $$\text{PDE } u_t = u_{xx}, \quad 0 < x < 1, \ 0 < t < \infty$$
>
> $$\text{BC } \begin{cases} u(0,t) = 0 \\ u(1,t) = \cos t \end{cases} \quad 0 < t < \infty$$
>
> $$\text{IC } u(x,0) = x, \ 0 \le x \le 1$$
>
> by
>
> a) transforming the problem to one with zero BCs
>
> b) solving the transformed problem by expanding the solution in terms of eigenfunctions.

Solution: Trying a solution of the form

$$u(x,t) = x\cos t + U(x,t)$$

we see that $U(x,t)$ satisfies

$$U_t = U_{xx} + x\sin t$$
$$U(0,t) = 0$$
$$U(1,t) = 0$$
$$U(x,0) = 0$$

We now expand the RHS

$$x\sin t = f_1(t)\sin(\pi x) + f_2(t)\sin(2\pi x) + \cdots$$

from which we find $f_n(t)$ to be

$$f_n(t) = 2\int_0^1 x\sin(t)\sin(n\pi x)\,dx = (-1)^{n+1}\left(\frac{2}{n\pi}\right)\sin(t), \ n = 1, 2, \ldots$$

We now can find the functions $T_n(t)$ which are the solutions of the initial-value problems

$$T_n' + (n\pi)^2 T_n = (-1)^{n+1}\left(\frac{2}{n\pi}\right)\sin(t)$$

$$T_n(0) = 0$$

which has solutions

$$T_n(t) = (-1)^{(n+1)}\frac{2}{n\pi}\int_0^1 e^{-(n\pi)^2(t-\tau)}\sin\tau\,d\tau$$

$$= \frac{(-1)^{(n+1)}2\pi\left(e^{(-n\pi)^2 t} + (n\pi)^2\sin t - \cos t\right)}{\pi^4 n^5 + n}$$

Hence, the solution of the transformed IBVP is

$$U(x,t) = \sum_{n=1}^{\infty} T_n(t)\sin(n\pi x) = \sum_{n=1}^{\infty}\frac{(-1)^{(n+1)}2\pi\left(e^{(-n\pi)^2 t} + (n\pi)^2\sin t - \cos t\right)}{\pi^4 n^5 + n}\sin(n\pi x)$$

and so the solution of the *original* IBVP is

$$u(x,t) = x\cos t + U(x,t)$$

$$= x\cos t + \sum_{n=1}^{\infty}\frac{(-1)^{(n+1)}2\pi\left(e^{(-n\pi)^2 t} + (n\pi)^2\sin t - \cos t\right)}{\pi^4 n^5 + n}\sin(n\pi x)$$

$$\Sigma\Omega\amalg\Delta\upsilon$$

Lesson 10: Integral Transforms (Sine and Cosine Transforms)

> 1. Prove the identities
>
> 1. $\Im_s[f'] = -\omega \Im_c[f]$
> 2. $\Im_s[f''] = \dfrac{2}{\pi}\omega f(0) - \omega^2 \Im_s[f]$
> 3. $\Im_c[f'] = -\dfrac{2}{\pi} f(0) + \omega \Im_s[f]$
> 4. $\Im_c[f''] = -\dfrac{2}{\pi} f'(0) - \omega^2 \Im_c[f]$

Solution:

1. Using the definition of the Sine transform and using integration by parts, we get

$$\Im_s[f'] = \frac{2}{\pi}\int_0^\infty f'(t)\sin(\omega t)\,dt$$

$$= \frac{2}{\pi}\left[f(t)\sin(\omega t)\right]_0^\infty - \omega\left(\frac{2}{\pi}\int_0^\infty f(t)\cos(\omega t)\,dt\right)$$

$$= -\omega\left(\frac{2}{\pi}\int_0^\infty f(t)\cos(\omega t)\,dt\right)$$

$$= -\omega \Im_c[f]$$

In the above integration we have used the fact that we must assume $f(t) \to 0$ as $t \to \infty$ else the transform will not exist.

To find $\Im_c[f']$, we compute

Lesson 10: Integral Transforms (Sine and Cosine Transforms)

$$\Im_c[f'] = \frac{2}{\pi}\int_0^\infty f'(t)\cos(\omega t)\,dt$$

$$= \frac{2}{\pi}\big[f(t)\cos(\omega t)\big]_0^\infty + \omega\left(\frac{2}{\pi}\int_0^\infty f(t)\sin(\omega t)\,dt\right)$$

$$= -\frac{2}{\pi}f(0) + \omega\left(\frac{2}{\pi}\int_0^\infty f(t)\sin(\omega t)\,dt\right)$$

$$= -\frac{2}{\pi}f(0) + \omega\Im_s[f]$$

To find $\Im_s[f'']$ and $\Im_c[f'']$, we simply use the results of the Sine and Cosine transform transforms of the first derivatives. That is

$$\Im_s[f''] = -\omega\Im_c[f']$$

$$= -\omega\left(-\frac{2}{\pi}f(0) + \omega\Im_s[f]\right)$$

$$= \frac{2}{\pi}\omega f(0) - \omega^2\Im_s[f]$$

Finally

$$\Im_c[f''] = -\frac{2}{\pi}f'(0) + \omega\Im_s[f']$$

$$= -\frac{2}{\pi}f'(0) + \omega\big(-\omega\Im_c[f]\big)$$

$$= -\frac{2}{\pi}f'(0) - \omega^2\Im_c[f]$$

---

2. Solve the initial-value problem

$$\text{ODE}:\quad \frac{dU}{dt} + (\alpha\omega)^2 U = \frac{2A\omega\alpha^2}{\pi}$$

$$\text{IC}:\quad U(0) = 0$$

Lesson 10: Integral Transforms (Sine and Cosine Transforms)

Solution: This problem is a simple review of what you learned in ordinary differential equations. The general solution of the ODE $y' + ay = b$ where a,b are constants is

$$y(t) = ce^{-at} + \frac{b}{a}$$

where c is an arbitrary constant. Plugging this general solution in the IC $y(0) = 0$ gives $c = -b/a$, hence the solution of the IVP

ODE: $y' + ay = b$
IC: $y(0) = 0$

is

$$y(t) = -\frac{b}{a}e^{-at} + \frac{b}{a} = \frac{b}{a}\left(1 - e^{-at}\right)$$

In our problem we have $a = (\alpha\omega)^2$ and $b = \dfrac{2A\omega\alpha^2}{\pi}$ and so the solution of the given problem is

$$U(t) = \frac{2A\omega\alpha^2}{\pi(\alpha\omega)^2}\left(1 - e^{-(\alpha\omega)^2 t}\right) = \frac{2A}{\pi\omega}\left(1 - e^{-(\alpha\omega)^2 t}\right)$$

We will now show you how to solve the following related problem using the Sine transform when the BC is changed to $u(0,t) = 0$.

Problem 3. Solve the following IBVP.

PDE $u_t = \alpha^2 u_{xx}$, $0 < x < \infty$, $0 < t < \infty$
BC $u_x(0,t) = 0$ $0 < t < \infty$
IC $u(x,0) = H(1-x)$, $0 \le x < \infty$

where $H(x)$ is the Heaviside (off-on) function defined by

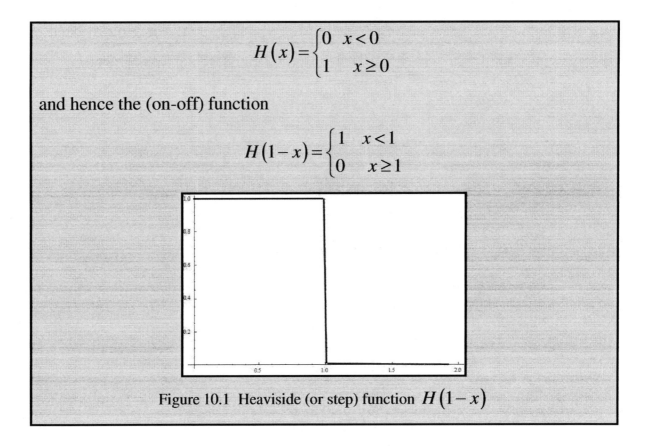

$$H(x) = \begin{cases} 0 & x < 0 \\ 1 & x \geq 0 \end{cases}$$

and hence the (on-off) function

$$H(1-x) = \begin{cases} 1 & x < 1 \\ 0 & x \geq 1 \end{cases}$$

Figure 10.1 Heaviside (or step) function $H(1-x)$

Solution: Since the BC is a derivative $u_x(0,t) = 0 \quad 0 < t < \infty$ we will use the Cosine transform

$$U(\omega,t) = \Im_c [u(x,t)] = \frac{2}{\pi} \int_0^\infty u(x,t) \sin(\omega x) dx$$

and transform the variable x, and using the general identity

$$\Im_c [f''(x)] = -\omega^2 \Im_c [f(x)] - \frac{2}{\pi} f'(0)$$

applied to the PDE and IC, we get the following ODE with IC:

Lesson 10: Integral Transforms (Sine and Cosine Transforms)

$$\text{ODE:} \quad \frac{dU}{dt}(t) = \alpha^2 \left[-\omega^2 U(t) - \frac{2}{\pi} u_x(0,t) \right]$$

$$= -(\alpha\omega)^2 U(t)$$

$$\text{IC:} \quad U(0) = \frac{2}{\pi} \int_0^\infty H(1-x) \cos(\omega x) dx$$

$$= \frac{2}{\pi} \int_0^1 \cos(\omega x) dx$$

$$= \frac{2}{\pi} \left[\frac{\sin(\omega x)}{\omega} \bigg|_0^1 \right]$$

$$= \frac{2}{\pi} \left[\frac{\sin(\omega)}{\omega} \right]$$

which simplifies to

$$\text{ODE:} \quad \frac{dU}{dt} + (\alpha\omega)^2 U(t) = 0$$

$$\text{IC:} \quad U(0) = \frac{2}{\pi} \left[\frac{\sin(\omega)}{\omega} \right]$$

and has the solution

$$U(\omega,t) = \frac{2}{\pi} \left[\frac{\sin(\omega)}{\omega} \right] e^{-(\alpha\omega)^2 t}$$

Taking the inverse Cosine transform, we get the solution

$$u(x,t) = \mathfrak{S}_c^{-1} \left[\frac{2}{\pi} \left[\frac{\sin(\omega)}{\omega} \right] e^{-(\alpha\omega)^2 t} \right]$$

$$= \frac{2}{\pi} \mathfrak{S}_c^{-1} \left[\left[\frac{\sin(\omega)}{\omega} \right] e^{-(\alpha^2 t)\omega^2} \right]$$

$$= \frac{1}{\sqrt{2\pi}} \left[\text{erf}\left(\frac{1-x}{2\alpha\sqrt{t}} \right) + \text{erf}\left(\frac{x+1}{2\alpha\sqrt{t}} \right) \right]$$

Lesson 10: Integral Transforms (Sine and Cosine Transforms)

The above solution may not be in familiar terms to some readers, but it can be evaluated accurately for any values of x and t. Below we have plotted the solution as a function of x for different values of time t. We see that the heat is diffusing out across the rod for larger and larger x. Heat is being lost from the rod at the left end point $x=0$. The steady state temperature of the rod is $u(x,\infty)=0$ for all x. The above integral for the inverse Cosine transform was done on Mathematica with the instruction

InverseFourierCosTransform[(2/pi)*(Sin[w]/w)*Exp[-a*w^2],w,x]

getting the result (printed format not always the best)

$$(\sqrt{\tfrac{\pi}{2}}\,(\mathrm{erf}((1-x)/(2\sqrt{a})) + \mathrm{erf}((x+1)/(2\sqrt{a}))))/\mathrm{pi}$$

where Mathematica's a is our $a = \alpha\sqrt{t}$. Graphs of the solution $u(x,t)$ as a function of x for different values of t are shown below. In all cases we have chosen $\alpha = 1$.

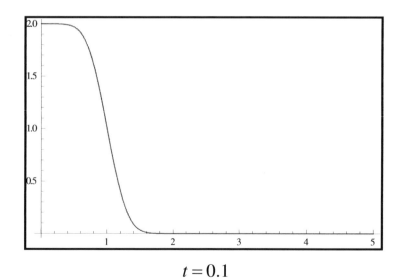

$t = 0.1$

Lesson 10: Integral Transforms (Sine and Cosine Transforms)

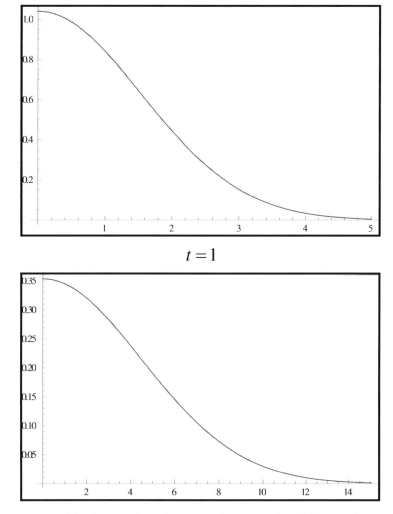

$t = 1$

$t = 10$ (note the time scale on x is different)

For comparison purposes, we now solve the problem where we change the BC from the Neumann condition $u_x(0,t) = 0$ to the Dirichlet condition $u(0,t) = 0$.

(Modified Problem 3) Solve the following IBVP.

Lesson 10: Integral Transforms (Sine and Cosine Transforms)

PDE $u_t = \alpha^2 u_{xx}$, $0 < x < \infty$, $0 < t < \infty$

BC $u(0,t) = 0$ $0 < t < \infty$

IC $u(x,0) = H(1-x)$, $0 \leq x < \infty$

where $H(x)$ is the Heaviside (off-on) function defined by

$$H(x) = \begin{cases} 0 & x < 0 \\ 1 & x \geq 0 \end{cases}$$

and hence the (on-off) function

$$H(1-x) = \begin{cases} 1 & x < 1 \\ 0 & x \geq 1 \end{cases}$$

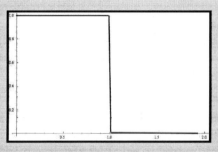

Figure 11.1 Heaviside (or step) function $H(1-x)$

Solution: Using the Sine transform

$$\mathcal{F}_s[u] = \frac{2}{\pi} \int_0^\infty u(x,t) \sin(\omega x)\, dx$$

we get

$$\mathcal{F}_s[u_t] = \frac{2}{\pi} \int_0^\infty u_t(x,t) \sin(\omega x)\, dx = \frac{\partial}{\partial t} \mathcal{F}_s[u] = \frac{d}{dt} U(t)$$

$$\mathcal{F}_s[u_{xx}] = \frac{2}{\pi} \int_0^\infty u_{xx}(x,t) \sin(\omega x)\, dx = \frac{2}{\pi} \omega u(0,t) - \omega^2 U = -\omega^2 U(t)$$

Hence the PDE in x and t is now transformed to a first-order ODE in t (the variable x in the PDE is transformed to ω, which now is considered to be a parameter in the ODE in t). The IC

$$\text{IC} \quad u(x,0) = H(1-x), \quad 0 \leq x < \infty$$

in the original IBVP is transformed to the constant (constant insofar as the ODE in t is concerned)

$$U(0) = 2\int_0^\infty H(x)\sin(\omega x)\,dx = 2\int_0^1 \sin(\omega x)\,dx = 2\left(\frac{1-\cos\omega}{\omega}\right)$$

Hence, we now have the IVP problem in variable t

$$\text{ODE:} \quad \frac{dU}{dt} + \omega^2 U = 0$$

$$\text{IC:} \quad U(0) = 2\left(\frac{1-\cos\omega}{\omega}\right)$$

which can easily be solved and found to be

$$U(t) = 2\left(\frac{1-\cos\omega}{\omega}\right) e^{-\omega^2 t}$$

Finally, taking the inverse Sine transform, gives

$$u(x,t) = 2\int_0^\infty \left(\frac{1-\cos\omega}{\omega}\right) e^{-\omega^2 t} \sin(\omega x)\,d\omega$$

$$= \sqrt{\frac{\pi}{2}}\left\{\text{erf}\left(\frac{1-x}{2\sqrt{t}}\right) + 2\,\text{erf}\left(\frac{x}{2\sqrt{t}}\right) - \text{erf}\left(\frac{x+1}{2\sqrt{t}}\right)\right\}$$

where *erf* and *erfc* are the error and complementary error functions, respectively.

The *Mathematica* statement for finding the above inverse Fourier sine transform is

Lesson 10: Integral Transforms (Sine and Cosine Transforms)

```
InverseFourierSinTransform[2*((1-Cos[w])/w)*Exp[-
w*w*t], w,x]
```

The solution may not be in familiar terms to some readers, but it can be evaluated accurately for any values of x and t. Below we have plotted the solution as a function of x for different values of time t. We see that the heat is diffusing out across the rod for larger and larger x. Heat is being lost from the rod at the left end point $x=0$. The steady state temperature of the rod is $u(x,\infty)=0$ for all x. Note that when $t=10$ in Figure 10.1 c) the temperature is nearly 0.

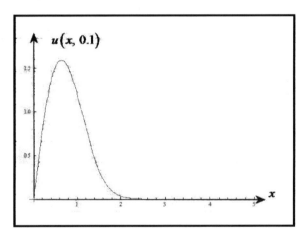

Figure 10,1 a) Solution $u(x,t)$ as a function of x when $t=0.1$

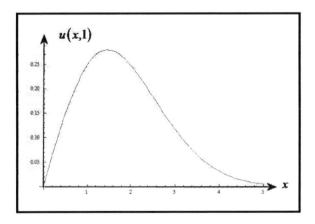

Figure 10,1b) Solution $u(x,t)$ as a function of x when $t=1$

Lesson 10: Integral Transforms (Sine and Cosine Transforms) 69

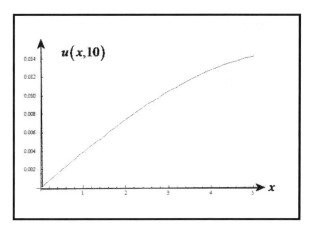

Figure 10,1 c) Solution $u(x,t)$ as a function of x when $t=10$

Note: It is useful to note that you could *not* solve the PDE

$$\text{PDE } u_t = \alpha^2 u_{xx} - V u_x, \quad 0 < x < \infty, \; 0 < t < \infty$$

that contains the first space derivative u_x by the Sine Transform, since the Sine transform of the *first derivative* involves the Cosine Transform. Also note that you could not transform the variable x with respect to the Laplace transform since the PDE contains a second derivative in x and there is only one side condition $u(0,t) = A$ at $x = 0$ whereas the Laplace transform would require two (and don't get the bright idea of adding another one like a flux $u_x(0,t) = B$ since requiring both temperature u and flux u_x at the boundary $x = 0$ does not make physical sense).

Second Note: If the BC at $x = 0$ were a flux condition $u_x(0,t) = \phi(t)$ then the Cosine Transform would be the one to use since the transform

$$\mathfrak{I}_c\left[u_{xx}(x,t)\right] = -\omega^2 \mathfrak{I}_c\left[u(x,t)\right] - \frac{2}{\pi} u_x(0,t)$$

involves the derivative at $x = 0$

Third Note: You many now want to compare separation of variables versus integral transforms. We only use separation of variables when the space domain is bounded; i.e. like $0 \leq x \leq 1$, whereas we can use the Sine (or Cosine) Transform when $0 < x < \infty$. Also, we cannot apply separation of variables for non homogeneous PDEs, whereas we can apply the Sine (or Cosine) transform. Also, separation of variables works for (linear of course) PDEs with variable coefficients but integral transform methods applies only to PDEs with constant coefficients (for the most part). Another point, separation of variables generally applies to PDEs defined on bounded regons; i.e. like $0 \leq x \leq 1$, where one arrives at infinite series solutions, whereas when one solves PDEs with integral transforms, one arrives at solutions in the form of integrals involving BCs and ICs of the problem, which may or may not be evaluated in terms of simple functions.

$$\Sigma \amalg \Delta \upsilon \Omega$$

Lesson 11 : The Fourier Series and Transform

1. What is the Fourier series expansion of the square wave

$$f(x) = \begin{cases} -1 & -1 < x < 0 \\ +1 & 0 < x < 1 \end{cases}$$

$f(x+2) = f(x)$ (periodic condition)

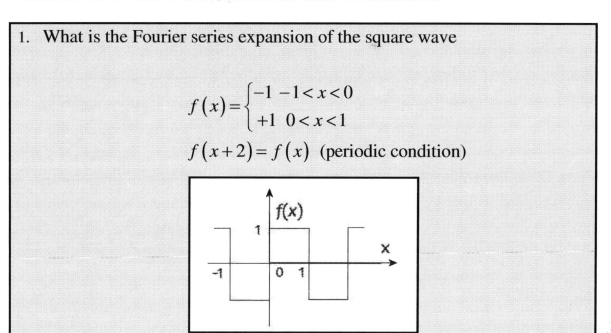

Solution: The function is odd, i.e. $f(x) = -f(-x)$ and so we know the Fourier series only contains sine terms, hence it has the form

$$S_f(x) = \sum_{n=1}^{\infty} a_n \sin(n\pi x)$$

(Note that the sine series starts with $n=1$ since $\sin(0)=0$.) To find the Fourier coefficients $a_n, n = 1, 2, \ldots$, we multiply each side of the equation by $\sin(m\pi x)$ and integrate each side of the equation from -1 to 1, and because of the orthogonality relation

$$\int_{-1}^{1} \sin(n\pi x) \sin(m\pi x) dx = \begin{cases} 0 & m \neq n \\ 1 & m = n \end{cases}$$

we get

$$\int_{-1}^{1} f(x) \sin(m\pi x) dx = \sum_{n=1}^{\infty} a_n \int_{-1}^{1} \sin(n\pi x) \sin(m\pi x) dx$$

$$= a_m \int_{-1}^{1} \sin^2(m\pi x) dx$$

$$= a_m$$

or

$$a_m = \int_{-1}^{1} f(x)\sin(m\pi x)dx$$
$$= -\int_{-1}^{0} \sin(m\pi x)dx + \int_{0}^{1} \sin(m\pi x)dx$$
$$= \int_{0}^{1} \sin(m\pi x)dx + \int_{0}^{1} \sin(m\pi x)dx$$
$$= 2\int_{0}^{1} \sin(m\pi x)dx$$
$$= -\frac{2}{m\pi}\left(\cos(m\pi x)\big|_{0}^{1}\right)$$
$$= \begin{cases} \dfrac{4}{m\pi} & m = 1,3,5,\ldots \\ 0, & m = 2,4,\ldots \end{cases}$$

and so the Fourier Sine series of f is

$$S_f(x) = \frac{4}{\pi}\sum_{n=1}^{\infty}\frac{1}{2n-1}\sin\left[(2n-1)\pi x\right]$$
$$= \frac{4}{\pi}\left[\sin(\pi x) + \frac{1}{3}\sin(3\pi x) + \frac{1}{5}\sin(5\pi x) + \cdots\right]$$

The first few terms of this series are shown in Figure 11.1.

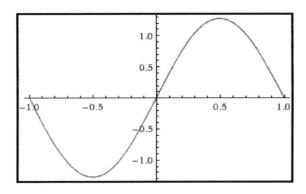

Figure 11.1 a) First term of the Fourier series

Lesson 11: The Fourier Series and Transform

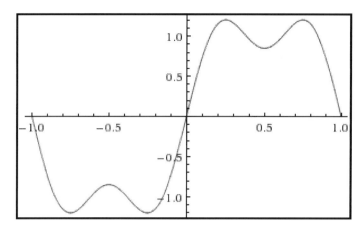

Figure 11.1 b) First 2 terms of the Fourier series

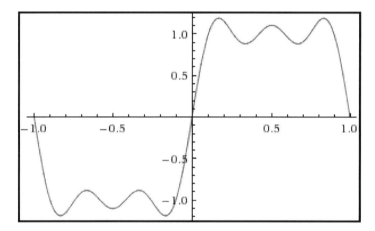

Figure 11.1 c) First 3 terms of the Fourier series

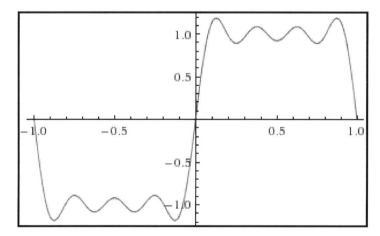

Figure 11.1 d) First 4 terms of the Fourier series

2. The sawtooth function

$$f(x) = x \quad -L < x \leq L$$

$$f(x+L) = f(x) \quad \text{(periodic condition)}$$

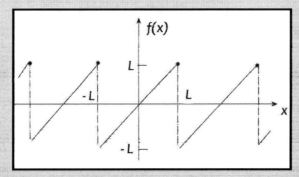

has a Fourier sine series given by

$$S_f(x) = \frac{2L}{\pi}\left[\sin\left(\frac{\pi x}{L}\right) - \frac{1}{2}\sin\left(\frac{2\pi x}{L}\right) + \frac{1}{3}\sin\left(\frac{3\pi x}{L}\right) - \cdots\right]$$

Show that if we differentiate this series term by term, we arrive at an infinite series that does not represent the derivative of the sawtooth function.

Solution : The Fourier sine series $S_f(x)$ represents a sawtooth curve, whose derivative is 1 except at the integers where the sawtooth function is not continuous, hence not derivative. If we differentiate the sine series $S_f(x)$ term by term, we do not get a series that represents the derivative of the sawtooth function since

$$\frac{dS_f(x)}{dx} = 2\left[\cos\left(\frac{\pi x}{L}\right) - \cos\left(\frac{2\pi x}{L}\right) + \cos\left(\frac{3\pi x}{L}\right) - \cdots\right]$$

doesn't represent anything since the series does not converge for any value of x.

Lesson 11: The Fourier Series and Transform

3. Graph the frequency spectrum of the following periodic functions.

   a) $f(x) = \sin x$

   b) $f(x) = \sin x + \cos 2x$

   c) $f(x) = \sin x + \cos x + \frac{1}{2} \sin 3x$

Solution : If the Fourier series $\mathcal{F}[f]$ of a function $f(x)$ is

$$\mathcal{F}[f] = \sum_{n=0}^{\infty} \left[a_n \cos(n\pi x) + b_n \sin(n\pi x) \right]$$

$$= a_0 + \sum_{n=1}^{\infty} \left[a_n \cos(n\pi x) + b_n \sin(n\pi x) \right]$$

then its frequency spectrum is the function C_n, $n = 0, 1, 2, ...$ (or sequence if you like), defined on the non-negative integers by

$$C_n = \sqrt{a_n^2 + b_n^2} , \quad n = 0, 1, 2, ...$$

(Note: b_0 is always 0 in a Fourier series) Computing these values for each function, we have the following graphs.

a) For $f(x) = \sin x$ we have $a_n = 0, n = 0, 1, 2, ...$ and $b_1 = 1, b_n = 0, n = 0, 2, 3, ...$ hence $C_n = \sqrt{a_n^2 + b_n^2}$, $n = 0, 1, 2, ...$ gives the values $C_1 = 1$ all the other values of C_n zero. The graph is shown in Figure 11.2a).

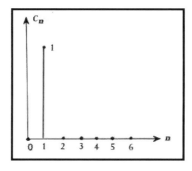

Figure 11.2 a) Frequency spectrum of $f(x) = \sin x$

b) For $f(x) = \sin x + \cos 2x$ we have $a_1 = 1$, $b_2 = 1$ all other values of a_n, b_n zero, hence $C_1 = 1$, $C_2 = 1$, all other values of C_n zero. The graph is shown in Figure 11.2b).

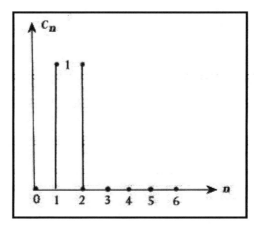

Figure 11.2 b) Frequency spectrum of $f(x) = \sin x + \cos 2x$

c) For $f(x) = \sin x + \cos x + \dfrac{1}{2}\sin 3x$ we have $a_1 = 1$, $b_1 = 1$, $b_3 = 1/2$ all other values of a_n, b_n zero, hence $C_1 = \sqrt{2}$, $C_3 = 1/2$, all other values of C_n zero. The graph is shown in Figure 11.2b).

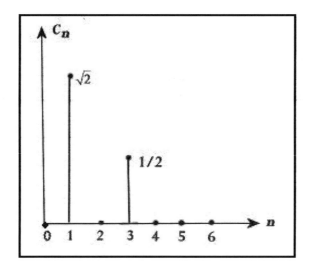

Figure 11.2 c) Frequency spectrum of $f(x) = \sin x + \cos x + \dfrac{1}{2}\sin 3x$

Lesson 11: The Fourier Series and Transform 77

4. What is the Fourier transform $F(\xi)$ and frequency spectrum $C(\xi) = |F(\xi)|$ of the function

$$f(x) = \begin{cases} 1 & -1 < x < 1 \\ 0 & \text{elsewhere} \end{cases}$$

Solution: Computing the Fourier transform we find

$$F[f] = \frac{1}{\sqrt{2\pi}} \int_{-\infty}^{\infty} e^{-i\xi x} f(x) \, dx$$

$$= \frac{1}{\sqrt{2\pi}} \int_{-1}^{1} e^{-i\xi x} \, dx$$

$$= \frac{1}{\sqrt{2\pi}} \left(\frac{-1}{i\xi}\right) e^{-i\xi x} \Big|_{-1}^{1}$$

$$= \frac{-\sqrt{2}}{\xi\sqrt{\pi}} \left[\frac{e^{i\xi} - e^{-i\xi}}{2i}\right]$$

$$= -\sqrt{\frac{2}{\pi}} \left(\frac{\sin \xi}{\xi}\right)$$

Since the transform is a real function, its spectrum is simply its absolute value

$$|F(\zeta)| = \sqrt{\frac{2}{\pi}} \left|\frac{\sin \xi}{\xi}\right|$$

5. Show that the absolute value of the complex function

$$F(\xi) = \frac{1}{1 + i\xi} \quad \text{is} \quad |F(\xi)| = \sqrt{\frac{1}{1 + \xi^2}}$$

Solution : The idea is to write the transform in the form $F(\xi) = a(\xi) + ib(\xi)$ and then compute $|F(\xi)| = \sqrt{a^2(\xi) + b^2(\xi)}$. In this problem, we do this by using the process of rationalizing the denominator, or

$$F(\xi) = \left(\frac{1}{1+i\xi}\right)\left(\frac{1-i\xi}{1-i\xi}\right) = \frac{1-i\xi}{1+\xi^2} = \left(\frac{1}{1+\xi^2}\right) + i\left(\frac{-\xi}{1+\xi^2}\right)$$

Hence

$$|F(\xi)| = \sqrt{\left(\frac{1}{1+\xi^2}\right)^2 + \left(\frac{-\xi}{1+\xi^2}\right)^2}$$

$$= \sqrt{\frac{1+\xi^2}{(1+\xi^2)^2}}$$

$$= \sqrt{\frac{1}{1+\xi^2}}$$

---

6. Verify the following orthogonality properties of the sin and cosine functions on the interval $[-L, L]$.

$$\int_{-L}^{L} \sin\left(\frac{m\pi x}{L}\right)\sin\left(\frac{n\pi x}{L}\right)dx = \begin{cases} 0 & m \neq n \\ L & m = n \end{cases}$$

$$\int_{-L}^{L} \cos\left(\frac{m\pi x}{L}\right)\cos\left(\frac{n\pi x}{L}\right)dx = \begin{cases} 0 & m \neq n \\ L & m = n \end{cases}$$

$$\int_{-L}^{L} \sin\left(\frac{m\pi x}{L}\right)\cos\left(\frac{n\pi x}{L}\right)dx = 0 \quad \text{all } m, n = 1, 2, 3, \ldots$$

Solution : Straight forward integration. Hint: Use the identities

Lesson 11: The Fourier Series and Transform

$$\sin\theta\sin\phi = \frac{1}{2}\left[\cos(\theta-\phi)-\cos(\theta+\phi)\right]$$

$$\cos\theta\cos\phi = \frac{1}{2}\left[\cos(\theta-\phi)+\cos(\theta+\phi)\right]$$

$$\sin\theta\cos\phi = \frac{1}{2}\left[\sin(\theta+\phi)+\sin(\theta-\phi)\right]$$

We will prove the orthogonality condition

$$\int_{-L}^{L}\sin\left(\frac{m\pi x}{L}\right)\cos\left(\frac{n\pi x}{L}\right)dx$$

$$= \frac{1}{2}\int_{-L}^{L}\left[\sin(m+n)\frac{\pi x}{L}+\sin(m-n)\frac{\pi x}{L}\right]dx$$

$$= \frac{1}{2}\left[\frac{-L}{(m+n)\pi}\cos(m+n)\frac{\pi x}{L}\bigg|_{-L}^{L}\right]+\frac{1}{2}\left[\frac{-L}{(m-n)\pi}\cos(m-n)\frac{\pi x}{L}\bigg|_{-L}^{L}\right]$$

$$= \frac{-L}{2(m+n)\pi}\left[\cos(m+n)\pi-\cos(m+n)\pi\right]$$

$$+\frac{-L}{2(m-n)\pi}\left[\left(\cos(m-n)\pi-\cos(m-n)\pi\right)\right]$$

$$= 0 \quad \text{all } m,n = 0,1,2,...$$

It is good to memorize the following popular orthogonality relations.

$$\int_{-1}^{1} \sin(m\pi x)\sin(n\pi x)\,dx = \begin{cases} 0 & m \neq n \\ 1 & m = n \end{cases}$$

$$\int_{-1}^{1} \cos(m\pi x)\cos(n\pi x)\,dx = \begin{cases} 0 & m \neq n \\ 1 & m = n \end{cases}$$

$$\int_{-L}^{L} \sin(m\pi x)\cos(n\pi x)\,dx = 0, \text{ all } m,n = 1,2,\ldots$$

$$\int_{-\pi}^{\pi} \sin(mx)\sin(nx)\,dx = \begin{cases} 0 & m \neq n \\ \pi & m = n \end{cases}$$

$$\int_{-\pi}^{\pi} \sin(mx)\cos(nx)\,dx = \begin{cases} 0 & m \neq n \\ \pi & m = n \end{cases}$$

$$\int_{-\pi}^{\pi} \sin(mx)\cos(nx)\,dx = 0, \text{ all } m,n = 1,2,\ldots$$

ΣΥΠΩV

Lesson 12 : The Fourier Transform and Its Applications to PDEs

> 1. Find the Fourier transform of
> $$f(x) = \begin{cases} 0 & x < 0 \\ e^{-x} & x \geq 0 \end{cases}$$

Solution From the definition of the Fourier transform, we have

$$F(\xi) = \frac{1}{\sqrt{2\pi}} \int_{-\infty}^{\infty} f(x) e^{-i\xi x} \, dx$$

$$= \frac{1}{\sqrt{2\pi}} \int_{0}^{\infty} e^{-x} e^{-i\xi x} \, dx$$

$$= \frac{1}{\sqrt{2\pi}} \int_{0}^{\infty} e^{-(1+i\xi)x} \, dx$$

$$= \frac{1}{\sqrt{2\pi}} \frac{-1}{1+i\xi} e^{-(1+i\xi)x} \Big|_{0}^{\infty}$$

$$= \frac{1}{\sqrt{2\pi}} \frac{1}{1+i\xi}$$

To find the Fourier spectrum, we first write

$$F(\xi) = \frac{1}{\sqrt{2\pi}} \frac{1}{1+i\xi} \frac{1-i\xi}{1-i\xi} = \frac{1}{\sqrt{2\pi}} \frac{1-i\xi}{1+\xi^2} = \frac{1}{\sqrt{2\pi}} \left[\left(\frac{1}{1+\xi^2} \right) + i \left(\frac{-\xi}{1+\xi^2} \right) \right]$$

Hence the Fourier spectrum is

$$|F(\xi)| = \frac{1}{\sqrt{2\pi}} \sqrt{\left(\frac{1}{1+\xi^2}\right)^2 + \left(\frac{-\xi}{1+\xi^2}\right)^2}$$

$$= \frac{1}{\sqrt{2\pi}} \sqrt{\frac{1+\xi^2}{(1+\xi^2)^2}}$$

$$= \frac{1}{\sqrt{2\pi}} \frac{1}{\sqrt{1+\xi^2}}$$

> **2.** Verify that the Fourier transform and inverse Fourier transform are linear transformations.

Solution: The calculus identity

$$\Im[c_1 f + c_2 g] = \frac{1}{\sqrt{2\pi}} \int_{-\infty}^{\infty} [c_1 f(x) + c_2 g(x)] e^{-i\xi x} \, dx$$

$$= c_1 \left\{ \frac{1}{\sqrt{2\pi}} \int_{-\infty}^{\infty} [f(x)] e^{-i\xi x} \, dx \right\} + c_2 \left\{ \frac{1}{\sqrt{2\pi}} \int_{-\infty}^{\infty} [g(x)] e^{-i\xi x} \, dx \right\}$$

$$= c_1 \Im[f] + c_2 \Im[g]$$

shows that the Fourier transform is a linear transform. Also, if $F(\xi)$ and $G(\xi)$ are the Fourier transforms of functions $f(x)$ and $g(x)$, respectively, then the inverse Fourier transform is also linear due to the following linear property of the integral:

$$\Im^{-1}[c_1 F + c_2 G] = \frac{1}{\sqrt{2\pi}} \int_{-\infty}^{\infty} [c_1 F(\xi) + c_2 G(\xi)] e^{i\xi x} \, d\xi$$

$$= c_1 \left\{ \frac{1}{\sqrt{2\pi}} \int_{-\infty}^{\infty} [F(\xi)] e^{i\xi x} \, d\xi \right\} + c_2 \left\{ \frac{1}{\sqrt{2\pi}} \int_{-\infty}^{\infty} [G(\xi)] e^{i\xi x} \, d\xi \right\}$$

$$= c_1 \Im^{-1}[F] + c_2 \Im^{-1}[G]$$

Lesson 12 : The Fourier Transform and Its Applications to PDE

> 3. Solve the following initial-value problem by the Fourier transform.
>
> PDE: $u_t = \alpha^2 u_{xx}$ $\quad -\infty < x < \infty, \ 0 < t < \infty$
>
> IC: $u(x,0) = e^{-x^2}$ $\quad -\infty < x < \infty$

Solution Taking the Fourier transform of the partial derivatives

$$\Im[u_t] = \frac{1}{\sqrt{2\pi}} \int_{-\infty}^{\infty} u_t(x,t) e^{-i\xi x} dx$$

$$= \frac{d}{dt}\left[\frac{1}{\sqrt{2\pi}} \int_{-\infty}^{\infty} u(x,t) e^{-i\xi x} dx\right]$$

$$= \frac{d}{dt}\Im[u]$$

$$= \frac{d}{dt}U(t)$$

$$\Im[u_{xx}] = \frac{1}{\sqrt{2\pi}} \int_{-\infty}^{\infty} u_{xx}(x,t) e^{-i\xi x} dx = -\xi^2 \Im(u) = -\xi^2 U(t)$$

and plugging these values in the PDE and IC, we get

$$\text{ODE:} \quad \frac{dU(t)}{dt} = -(\alpha\xi)^2 U(t) \quad 0 < t < \infty$$

$$\text{IC:} \quad U(0) = \Im\left[e^{-x^2}\right] = \frac{1}{\sqrt{2}} e^{-(\xi/2)^2}$$

Remembering your ODEs, we solve this IVP and get

$$U(t,\xi) = U(0) e^{-(\alpha\xi)^2 t} = \frac{1}{\sqrt{2}} e^{-(\xi/2)^2} e^{-(\alpha\xi)^2 t} \quad 0 < t < \infty$$

which has an inverse Fourier transform

$$u(x,t) = \Im^{-1}[U(\xi,t)]$$

$$= \frac{1}{\sqrt{2\pi}} \int_{-\infty}^{\infty} U(\xi,t) e^{i\xi x} d\xi$$

$$= \frac{1}{\sqrt{2\pi}} \int_{-\infty}^{\infty} \frac{1}{\sqrt{2}} e^{-(\xi/2)^2} e^{-(\alpha\xi)^2 t} e^{i\xi x} d\xi$$

$$= \frac{1}{2\sqrt{\pi}} \int_{-\infty}^{\infty} e^{-(\xi/2)^2} e^{-(\alpha\xi)^2 t} e^{i\xi x} d\xi$$

$$= \frac{1}{\sqrt{4\alpha^2 t + 1}} e^{-\left(\frac{x^2}{4\alpha^2 t + 1}\right)} \quad \text{(not so straight forward integration)}$$

Note: The above integral can be evaluated by collecting terms involving ξ in the exponent, completing the square, and then integrating. If not, you can always go to www.wolframalpha.com and enter

(1/(2*sqrt(pi)))* int exp(-(1/4)*z^2 - t*(a*z)^2 +I * z*x) dz z=-infinity..infinity

---

4. Verify the following properties of the Fourier transform

$$\Im[u_x] = i\xi \Im[u]$$
$$\Im[u_{xx}] = -\xi^2 \Im[u]$$

Solution From the definition of the Fourier transform and applying the integration by parts formula

$$\int_a^b u\,dv = uv \Big|_a^b - \int_a^b v\,du$$

we have

$$\Im[u_x] = \frac{1}{\sqrt{2\pi}} \int_{-\infty}^{\infty} u_x(x) e^{-i\xi x} dx$$

$$= u(x) e^{-i\xi x} \Big|_{-\infty}^{\infty} + i\xi \int_{-\infty}^{\infty} u(x) e^{-i\xi x} dx$$

$$= i\xi \Im[u]$$

Lesson 12 : The Fourier Transform and Its Applications to PDE

We can now apply the above formula to find

$$\Im[u_{xx}] = i\xi \Im[u_x] = i\xi[i\xi]\Im[u] = -\xi^2 \Im[u]$$

---

5. Verify that the convolution of two functions f and g can be written either as

$$(f*g)(x) = \frac{1}{\sqrt{2\pi}} \int_{-\infty}^{\infty} f(x-\xi)g(\xi)\,d\xi$$

or

$$(f*g)(x) = \frac{1}{\sqrt{2\pi}} \int_{-\infty}^{\infty} f(\xi)g(x-\xi)\,d\xi$$

Solution: Going from one convolution integral to the other is simply a change of variables. If we let $z = x - \xi$ in the first convolution integral, we get

$$(f*g)(x) = \frac{1}{\sqrt{2\pi}} \int_{-\infty}^{\infty} f(x-\xi)g(\xi)\,d\xi$$

$$= \frac{1}{\sqrt{2\pi}} \int_{\infty}^{-\infty} f(z)g(x-z)(-dz)$$

$$= \frac{1}{\sqrt{2\pi}} \int_{-\infty}^{\infty} f(z)g(x-z)\,dz$$

$$= \frac{1}{\sqrt{2\pi}} \int_{-\infty}^{\infty} f(\xi)g(x-\xi)\,d\xi$$

$$\Pi\nabla\Sigma\mho\nabla$$

Lesson 13: The Laplace Transform

> 1. Verify the following formula for the transform of the partial derivative
>
> $$L[u_t(x,t)] = sU(x,s) - u(x,0)$$

Solution: Using the definition of the Laplace transform and integration by parts (which will take the derivative off u_t and put it on e^{-st}, we have

$$L[u_t] = \int_0^\infty u_t(x,t) e^{-st}\, dt$$

$$= u(x,t) e^{-st} \Big|_{t=0}^{t=\infty} + s\int_0^\infty u(x,t) e^{-st}\, dt$$

$$= -u(x,0) + s\int_0^\infty u(x,t) e^{-st}\, dt$$

$$= sL[u(x,t)] - u(x,0)$$

> 2. Solve the following initial-value problem using the Laplace transform
>
> PDE: $u_t = \alpha^2 u_{xx}$ $-\infty < x < \infty,\ 0 < t < \infty$
>
> IC: $u(x,0) = \sin x$ $-\infty < x < \infty$

Solution: Since

$$L[u_t] = sU(s) - u(x,0)$$

$$L[u_{xx}] = \frac{d^2}{dx^2} U(s)$$

where we denote $U(s) \equiv U(x,s) = L[u(x,t)]$, the PDE, has independent variables x,t, and transforms to an ODE with independent variable x. The ODE also depends on the transformed variable s, but s is held constant in the ODE, so we treat it like a parameter. Hence, we have

Lesson 13: The Laplace Transform

$$\text{ODE:} \quad sU(x) - \sin x = \alpha^2 \frac{d^2 U(x)}{dx^2}$$

Any student who has studied ODEs should be able to solve this 2nd order, linear, nonhomogeneous ODE. If you can't solve this equation yourself, just go to www.wolframalpa.com and enter

$$a\wedge 2 * U''(x) - s * U(x) = \sin(x)$$

and wolframalpha will respond with

$$U(x,s) = c_1 e^{\sqrt{s}x/\alpha} + c_1 e^{-\sqrt{s}x/\alpha} - \frac{\sin x}{s + \alpha^2}$$

The inverse Laplace transform of this expression can be found in a table of inverse transforms or by use of computer program like *Mathematica* or *Maple*, and is

$$u(x,t) = L^{-1}\left[U(x,s)\right] = \sin x \, e^{-\alpha^2 t}, \quad -\infty < x < \infty, \ 0 \leq t < \infty$$

You can verify that this function satisfies both the PDE and ICs. You can also see that the solution is a 'decaying' sine curve. Does this solution make sense to you? It should, the energy gets spread out more and more on the real line.

---

3. Solve the following IBVP using the Laplace transform.

$$\text{PDE:} \quad u_t = u_{xx} \qquad 0 < x < \infty, \ 0 < t < \infty$$
$$\text{BC:} \quad u(0,t) = \sin t \qquad 0 < t < \infty$$
$$\text{IC:} \quad u(x,0) = 0 \qquad 0 < x < \infty$$

What is the physical interpretation of the problem?

Solution: Taking the Laplace transform (of the t variable) of this problem, we find

ODE: $sU(x,s) = \dfrac{d^2}{dx^2}U(x,s)$

IC: $U(0) = \dfrac{1}{s^2+1}$

Note that this IVP problem doesn't make sense inasmuch as the ODE is 2nd order and there is only one IC. We add a second condition from the physical fact we require the solution $U(x)$ to remain bounded for large x. If we solve the above ODE with the *single* IC we will get one arbitrary constant c_1 in the solution. Solving this problem with wolframalpha.com we enter

$$U''(x)=s*U(x), U(0)=1/(s^2+1)$$

and wolframalpha gives us

$$U(x) = \dfrac{e^{-\sqrt{s}x}\left(c_1 s^2 e^{2\sqrt{s}x} - c_1 s^2 + c_1 e^{2\sqrt{s}x} - c_1 + 1\right)}{s^2+1}$$

but the only bounded solution occurs when we pick $c_1 = 0$, and so we have the solution

$$U(x,s) = \dfrac{1}{s^2+1} e^{-x\sqrt{s}}$$

and finding the inverse Laplace transform, we get

Lesson 13: The Laplace Transform

$$u(x,t) = L^{-1}[U(x,s)]$$

$$= L^{-1}\left[\frac{1}{s^2+1}e^{-x\sqrt{s}}\right]$$

$$= L^{-1}\left[\frac{1}{s^2+1}\right] * L^{-1}\left[e^{-x\sqrt{s}}\right]$$

$$= \sin t * \frac{xe^{-x^2/4t}}{2\sqrt{\pi}\, t^{3/2}}$$

$$= \frac{1}{2\sqrt{\pi}} \int_0^t x \frac{e^{-x^2/4\xi}}{\xi^{3/2}} \sin(t-\xi)\, d\xi$$

The above convolution integral can be evaluated numerically for any (x,t) to essentially any accuracy you please. The above inverse transform of $e^{-x\sqrt{s}}$ was found with the help of wolframalpha by entering the command

inverse laplace transform exp(-x*sqrt(s)) which yields

$$\frac{x\, e^{-\frac{x^2}{4t}}}{2\sqrt{\pi}\, t^{3/2}} \quad \text{for } x > 0$$

You might try to imagine what the solution looks like. The temperature of the semi-infinite rod is initially zero, but we force the left end of the rod to have the periodic temperature $u(0,t) = \sin t$. How do these temperature look like as they move down the rod?

---

4. Solve the following two-point boundary value problem for $U(x)$ (treat s as a parameter; i.e. a constant).

ODE: $\quad \dfrac{d^2 U}{dx^2} - sU(x) = A, \quad 0 < x < 1$

BCs: $\quad \begin{cases} \dfrac{dU(0)}{dx} = 0 \\ U(1) = 0 \end{cases}$

Solution: As you learned in ODE, the general solution of the ODE is

$$U(x) = c_1 e^{\sqrt{s}\,x} + c_2 e^{-\sqrt{s}\,x} - \frac{A}{s}$$

where c_1, c_2 are arbitrary constants. To find c_1, c_2 we plug the general solution in the BCs, getting

$$\text{BCs}: \begin{cases} \dfrac{dU(0)}{dx} = c_1\sqrt{s} - c_2\sqrt{s} = 0 \\[2mm] U(1) = c_1 e^{\sqrt{s}} + c_2 e^{-\sqrt{s}} = \dfrac{A}{s} \end{cases}$$

which yields the constants

$$c_1 = c_2 = \frac{A}{s\left(e^{\sqrt{s}} + e^{-\sqrt{s}}\right)}$$

Plugging these values into the general solution $U(x)$ and simplifying, we get

$$U(x,s) = \frac{A}{s}\left(\frac{e^{\sqrt{s}\,x} + e^{-\sqrt{s}\,x}}{e^{\sqrt{s}} + e^{-\sqrt{s}}}\right) - \frac{A}{s} = \frac{A}{s}\left(\frac{\sinh\sqrt{s}\,x}{\sinh\sqrt{s}}\right) - \frac{A}{s}$$

$$\Sigma\Omega\Pi\amalg\sigma$$

Lesson 14: Duhamel's Principle

> Jean-Marie Duhamel (1797-1872) was a French mathematician and physicist who obtained a method of finding the solution of one PDE in terms of another.

1. Write the solution of the IBVP

$$\text{PDE: } u_t = u_{xx} \quad 0 < x < 1, \ 0 < t < \infty$$

$$\text{BC: } \begin{cases} u(0,t) = 0 \\ u(1,t) = f(t) \end{cases} \quad 0 < t < \infty$$

$$\text{IC: } u(x,0) = 0 \quad 0 < x < 1$$

in terms of the solution of the IBVP

$$\text{PDE: } w_t = w_{xx} \quad 0 < x < 1, \ 0 < t < \infty$$

$$\text{BC: } \begin{cases} w(0,t) = 0 \\ w(1,t) = \delta(t) \end{cases} \quad 0 < t < \infty$$

$$\text{IC: } w(x,0) = 0 \quad 0 < x < 1$$

Solution: Taking the Laplace transform of

$$\text{PDE: } u_t = u_{xx} \quad 0 < x < 1, \ 0 < t < \infty$$

$$\text{BC: } \begin{cases} u(0,t) = 0 \\ u(1,t) = f(t) \end{cases} \quad 0 < t < \infty$$

$$\text{IC: } u(x,0) = 0 \quad 0 < x < 1$$

and

$$\text{PDE: } w_t = w_{xx} \quad 0 < x < 1, \ 0 < t < \infty$$

$$\text{BC: } \begin{cases} w(0,t) = 0 \\ w(1,t) = \delta(t) \end{cases} \quad 0 < t < \infty$$

$$\text{IC: } w(x,0) = 0 \quad 0 < x < 1$$

we find

Lesson 14: Duhamel's Principle

$$\frac{d^2U(x)}{dx^2} - sU(s) = 0 \quad \bigg| \quad \frac{d^2W(x)}{dx^2} - sW(s) = 0$$
$$U(0) = 0 \quad \bigg| \quad W(0) = 0$$
$$U(1) = F(s) \quad \bigg| \quad W(1) = 1$$

which give solutions

$$U(x,s) = F(s) \left[\frac{\sinh \sqrt{s}\, x}{\sinh \sqrt{s}} \right] \quad W(x,s) = \left[\frac{\sinh \sqrt{s}\, x}{\sinh \sqrt{s}} \right]$$

Hence we have (watch carefully)

$$u(x,t) = \mathcal{L}^{-1}\left[F(s) \frac{\sinh \sqrt{s}\, x}{\sinh \sqrt{s}} \right]$$

$$= \mathcal{L}^{-1}[F(s)] * \mathcal{L}^{-1}\left[\left[\frac{\sinh \sqrt{s}\, x}{\sinh \sqrt{s}} \right]\right]$$

$$= w(x,t) * f(t)$$

2. Find the solution of the boundary value problem

ODE: $\dfrac{d^2U(x)}{dx^2} - sU(s) = 0$

BCs $\begin{cases} U(0) = 0 \\ U(1) = F(s) \end{cases}$

Solution: The ODE has the general solution

$$U(x) = c_1 e^{\sqrt{s}\, x} + c_2 e^{-\sqrt{s}\, x}$$

Lesson 14: Duhamel's Principle

To find the constants c_1, c_2 we plug the general solution in the BCs, getting

$$c_1 + c_2 = 0$$
$$c_1 e^{\sqrt{s}} + c_2 e^{-\sqrt{s}} = F(s)$$

and solving for c_1, c_2 gives

$$c_1 = \frac{F(s)}{e^{\sqrt{s}} - e^{-\sqrt{s}}}, \quad c_2 = -\frac{F(s)}{e^{\sqrt{s}} - e^{-\sqrt{s}}},$$

Hence, the solution is

$$U(x) = c_1 e^{\sqrt{s}x} + c_2 e^{-\sqrt{s}x}$$
$$= \frac{F(s)}{e^{\sqrt{s}} - e^{-\sqrt{s}}} \left[e^{\sqrt{s}x} - e^{-\sqrt{s}x} \right]$$
$$= F(x) \left[\frac{e^{\sqrt{s}x} - e^{-\sqrt{s}x}}{e^{\sqrt{s}} - e^{-\sqrt{s}}} \right]$$
$$= F(s) \left[\frac{\sinh \sqrt{s}\, x}{\sinh \sqrt{s}} \right]$$

---

3 What is the solution $u(x,t)$ of the IBVP

PDE: $u_t = u_{xx}$ $0 < x < 1, \ 0 < t < \infty$

BC: $\begin{cases} u(0,t) = 0 \\ u(1,t) = f(t) \end{cases}$ $0 < t < \infty$

IC: $u(x,0) = 0$ $0 \leq x \leq 1$

in terms of the solution $w(x,t)$ of

$$\begin{aligned}
&\text{PDE:} \quad w_t = w_{xx} \quad\quad 0 < x < 1,\ 0 < t < \infty \\
&\text{BC:} \quad \begin{cases} w(0,t) = 0 \\ w(1,t) = g(t) \end{cases} \quad 0 < t < \infty \\
&\text{IC:} \quad w(x,0) = 0 \quad\quad 0 \leq x \leq 1
\end{aligned}$$

Solution: Taking the Laplace transform of each problem gives

$$\begin{array}{|c|c|}
\hline
\dfrac{d^2 U(x)}{dx^2} - sU(s) = 0 & \dfrac{d^2 W(x)}{dx^2} - sW(s) = 0 \\
U(0) = 0 & W(0) = 0 \\
U(1) = F(s) & W(1) = G(s) \\
\hline
\end{array}$$

which give solutions

$$\boxed{U(x,s) = F(s)\left[\dfrac{\sinh\sqrt{s}\,x}{\sinh\sqrt{s}}\right]} \quad \boxed{W(x,s) = G(s)\left[\dfrac{\sinh\sqrt{s}\,x}{\sinh\sqrt{s}}\right]}$$

Hence, we can write

$$U(x,s) = F(s)\left[\dfrac{\sinh\sqrt{s}\,x}{\sinh\sqrt{s}}\right]$$

$$= \dfrac{F(s)}{G(s)} G(s)\left[\dfrac{\sinh\sqrt{s}\,x}{\sinh\sqrt{s}}\right]$$

Hence, we have

$$u(x,t) = \mathcal{L}^{-1}\left[\dfrac{F(s)}{G(s)}\right] * \mathcal{L}^{-1}\left[G(s)\left[\dfrac{\sinh\sqrt{s}\,x}{\sinh\sqrt{s}}\right]\right]$$

$$= \mathcal{L}^{-1}\left[\dfrac{F(s)}{G(s)}\right] * w(x,t)$$

Lesson 14: Duhamel's Principle

4. Use the result in Problem 3 to find the solution of

$$\text{PDE: } u_t = u_{xx} \qquad 0 < x < 1, \ 0 < t < \infty$$

$$\text{BC: } \begin{cases} u(0,t) = 0 \\ u(1,t) = \sin t \end{cases} \qquad 0 < t < \infty$$

$$\text{IC: } u(x,0) = 0 \qquad 0 \le x \le 1$$

in terms of the solution $w(x,t)$ of

$$\text{PDE: } w_t = w_{xx} \qquad 0 < x < 1, \ 0 < t < \infty$$

$$\text{BC: } \begin{cases} w(0,t) = 0 \\ w(1,t) = 1 \end{cases} \qquad 0 < t < \infty$$

$$\text{IC: } w(x,0) = 0 \qquad 0 \le x \le 1$$

Solution: Here

$$f(t) = \sin t \Rightarrow F(s) = 1/(s^2 + 1)$$
$$g(t) = 1 \Rightarrow G(s) = 1/s$$

and so

$$\frac{F(s)}{G(s)} = \frac{s}{s^2 + 1} \Rightarrow \mathcal{L}^{-1}\left[\frac{s}{s^2 + 1}\right] = \sin t$$

Hence

$$u(x,t) = \mathcal{L}^{-1}\left[\frac{F(s)}{G(s)}\right] * w(x,t)$$

$$= \sin t * w(x,t)$$

$$= \int_0^t w(x, t - \xi) \sin \xi \, d\xi$$

This not part of the problem, but we could now observe that the steady state for $w(x,t)$ witll be $w(x,\infty) = x$ and so letting

$$w(x,t) = x + U(x,t)$$

and arrive at the IBVP for the transient $U(x,t)$:

PDE: $U_t = U_{xx}$ $\quad 0 < x < 1,\ 0 < t < \infty$

BC: $\begin{cases} U(0,t) = 0 \\ U(1,t) = 0 \end{cases}$ $\quad 0 < t < \infty$

IC: $U(x,0) = -x$ $\quad 0 \leq x \leq 1$

which we can solve by separation of variables. Hence we have $U(x,t)$ and thus $w(x,t)$ and thus $u(x,t)$. Although $w(x,t)$ is an infinite series, we can use the first 4 or 5 terms to get an approximation of $u(x,t)$.

$$\Sigma\nabla\Omega\sigma\Pi$$

Lesson 15 : The Convection Term u_x in Diffusion Problems

> 1. Solve the initial-value problem
>
> $$\text{PDE: } u_t = u_{xx} - u_x \qquad -\infty < x < \infty, \ 0 < t < \infty$$
> $$\text{IC: } u(x,0) = \sin x \qquad -\infty < x < \infty$$

Solution: We saw in Lesson 8 that the transformation

$$u(x,t) = e^{\frac{1}{\alpha^2}\left(\frac{vx}{2} - \frac{v^2 t}{4}\right)} U(x,t)$$

transforms the PDE

$$u_t = \alpha^2 u_{xx} - v u_x$$

in dependent variable $u(x,t)$ to the simpler PDE

$$U_t = U_{xx}$$

in dependent variable $U(x,t)$. But there is another way to eliminate the convection term $-u_x$ (we let $v=1$ as in the problem) and that is by transforming the *independent* variables x,t to two new ones ξ, τ via the transformation

$$\tau = t$$
$$\xi = x - t$$

(We haven't really changed the time variable t, but it is more convenient to give it a new name, which we call τ). To find the new PDE with new independent variables ξ, τ we make use of the diagram in Figure 15.1 to find the following partial derivatives.

$$u_t = u_\xi \xi_t + u_\tau \tau_t = -u_\xi + u_\tau$$
$$u_x = u_\xi \xi_x + u_\tau \tau_x = u_\xi$$
$$u_{xx} = (u_x)_x = (u_\xi)_x = u_{\xi\xi} \xi_x + u_{\xi\tau} \tau_x = u_{\xi\xi}$$

Lesson 15: The Convection Term u_x in Diffusion Problems

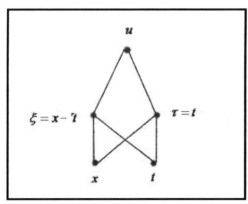

Figure 15.1 Functional relationship of variables

Plugging these derivatives in the original PDE, we find

$$-u_\xi + u_\tau = u_{\xi\xi} - u_\xi$$

or the new problem

$$u_\tau = u_{xx}$$
$$u(\xi,0) = \sin \xi$$

which has the solution

$$u(\xi,\tau) = e^{-\xi} \sin \xi$$

Plugging back in for the original variables gives the solution of the original problem to be

$$u(z,t) = e^{-t} \sin(x-t)$$

---

2. Solve the following diffusion convection equation

$$\text{PDE: } u_t = u_{xx} - 2u_x \qquad -\infty < x < \infty, \ 0 < t < \infty$$

$$\text{IC: } u(x,0) = e^x \sin x \qquad -\infty < x < \infty$$

by making a transformation like we did in Problem 1 in Lesson 8.

Solution: In Problem 1, Lesson 8 we showed the transformation that maps

$$u_t = \alpha^2 u_{xx} - v u_x \to U_t = \alpha^2 U_{xx}$$

Lesson 15: The Convection Term u_x in Diffusion Problems

is

$$u(x,t) = e^{\frac{1}{\alpha^2}\left(\frac{vx}{2} - \frac{v^2 t}{4}\right)} U(x,t)$$

Hence, in this problem we let

$$u(x,t) = e^{(x-t)} U(x,t)$$

hence

$$u_t = -e^{(x-t)} U(x,t) + e^{(x-t)} U_t(x,t)$$
$$u_x = e^{(x-t)} U(x,t) + e^{(x-t)} U_x(x,t)$$
$$u_{xx} = e^{(x-t)} U(x,t) + 2e^{(x-t)} U_x(x,t) + e^{(x-t)} U_{xx}(x,t)$$

Plugging these values in the equation

$$u_t = u_{xx} - 2u_x$$

after some cancelation gives $U_t = U_{xx}$. The initial condition

$$u(x,0) = e^x \sin x$$

then becomes

$$u(x,0) = e^x U(x,0) = e^x \sin x$$

or $U(x,0) = \sin x$. Hence, the transformed problem is

PDE: $U_t = U_{xx}$ $-\infty < x < \infty, \ 0 < t < \infty$
IC: $U(x,0) = \sin x$ $-\infty < x < \infty$

which has the solution

$$U(x,t) = e^{-t} \sin x$$

and thus the original problem has the solution

Lesson 15: The Convection Term u_x in Diffusion Problems

$$u(x,t) = e^{(x-t)} U(x,t)$$
$$= e^{(x-t)} e^{-t} \sin x$$
$$= e^{(x-2t)} \sin x$$

---

3. Solve

$$\text{PDE: } u_t = -2u_x \qquad -\infty < x < \infty, \ 0 < t < \infty$$
$$\text{IC: } u(x,0) = e^{-x^3} \qquad -\infty < x < \infty$$

Solution: In the next lesson, we will give a formal description of solving pure convection equations like this problem (no u_{xx} term only a term in u_x). For the time being, we can interpret this problem as describing the amount $u(x,t)$ of a substance moving to the right along the x-axis with velocity 2, we can imagine the initial wave $u(x,0) = e^{-x^3}$, we might be temped to try the solution

$$u(x,t) = e^{(x-3t)^2}$$

You can verify by direct substitution that this function satisfies both the PDE and the IC.

---

4. What is the solution of the convection problem

$$\text{PDE: } u_t = u_{xx} - Vu_x \qquad -\infty < x < \infty, \ 0 < t < \infty$$
$$\text{IC: } u(x,0) = e^{-x^2} \qquad -\infty < x < \infty$$

and check your solution. Hint: Note that the transformation to moving coordinates allows us to essentially neglect the convection term $-Vu_x$

Solution: Letting $\tau = t, \xi = x - Vt$, using the following diagram to help with our partial derivatives.

Lesson 15: The Convection Term u_x in Diffusion Problems

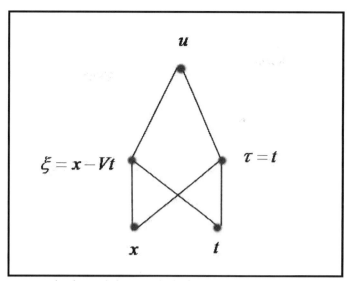

Diagram to help with partial derivatives of compositions

We find

$$u_x = u_\xi \xi_x + u_\tau \tau_x = u_\xi$$

$$u_{xx} = \frac{\partial}{\partial x}(u_\xi) = u_{\xi\xi}\xi_x + u_{\xi\tau}\tau_x = u_{\xi\xi}$$

$$u_t = u_\xi(-V) + u_\tau \tau_t = u_\tau - V u_\xi$$

Plugging these values in original IVP yields

PDE: $u_t = u_{\xi\xi}$ $\quad -\infty < \xi < \infty, \ 0 < t < \infty$

IC: $u(\xi, 0) = e^{-\xi^2}$ $\quad -\infty < \xi < \infty$

We then solve this problem for $u(\xi, \tau)$ and then substitute back $\tau = t, \xi = x - Vt$ to get u as a function of x and t.

$$\Sigma\amalg\nabla\Delta\mho$$

Ten Reasons for Not Naming Your Cat Calculus

I thank Peter Roget for letting me express my angst when I say the worst, lousy, rotten, dire, god-awful mistake I ever made was when I took my cat Calculus to college with me. I had Calculus from when I was a young boy and couldn't bear to part with him. Here are just a few problems I experienced during my college years with Calculus.

One weekend I called my girlfriend and asked her if she wanted to come over. She said yes. I told her we could spend the afternoon toying with Calculus. She never came and I never saw her again.

The next day, I took Calculus to the vet for some shots. On the way home, I met a friend who asked me how my classes were going. I told him everything was going great, my English professor was helping me with my writing, I had a tutor for my Spanish class, and I had just been to the vet who was helping me with Calculus.

A couple of weeks later, my drinking buddies called and asked if I wanted to go to a party that weekend. They said there would be lots of girls there. I told them sure and that I had just spent the better part of the day playing with Calculus. That was the last time they invited me.

In my off-time, I volunteered in a nursing home, assisting the elderly residents. I suggested to the manager that Calculus might be a fun diversion for the elderly, that even a 90-year old could spend quality time with Calculus. She told me to continue with the bedpans.

Later that semester, I dated a cheerleader and she asked me if I was well rounded and not just a nerdy math type. I told her I was well-rounded and liked many things, and that I knew Calculus since I was one year old. I never saw her after that.

At the end of a semester, I had a nervous breakdown and went to the college psychologist. He worried about me getting over-stressed and a little wacky and asked if I had a pet that might take my mind off my studies. I told him I had a cat and my neighbor had a dog who loved Calculus. The doctor doubled my prescription.

The doctor sent me to the campus infirmary for psychological evaluation, wanting to learn more about my early life. I told them I had a normal childhood even from the age of two, spending many happy hours with in my baby crib with Calculus. They gave me some more pills.

My advisor asked me at the end of my freshman year how I planned on spending the summer. I told him I wanted to write a children's book about Calculus.

By the time I was a senior, I gained somewhat of a reputation as a math whiz and tutor. One day I was contacted by a freshman who was worried about upcoming exams. I told her not to worry, that if she came to my residence I would help her. She said she didn't have much money for the tutoring, but said it was important that she and calculus were on good terms. I told her I wouldn't charge for the tutoring, but if she serious about being on the best of terms, it would be a good idea to bring along a tin of sardines and a bag of cat liter,

At the end of my college career I asked my advisor to write a letter of recommendation for me for grad school. He asked me what calculus had taught me in my four years of college. I told him Calculus taught me that I'd never get a dog. I never got into grad school.

After graduation from college I had an interview with a company. The interviewer asked how I might use mathematics in solving the company's problems. I told him I'd love to apply mathematics in solving their problems, and in fact Calculus might come in handy in rodent control in the company's warehouse. I never got the job.

$$\Sigma\Theta\Gamma\Xi\Omega\Upsilon$$

Section 3: Hyperbolic Type Problems

Lesson 16 : One Dimensional Wave Equation

1. Derive the transmission line equation

$$v_{xx} = CLv_{tt} + (CR + GL)v_t + GRv$$

for v from the following system of first-order PDEs

$$i_x + Cv_t + Gv = 0$$
$$v_x + Li_t + Ri = 0$$

Solution: If we differentiate the first equation with respect to t and the second equation with respect to x, we get

$$i_{xt} + Cv_{tt} + Gv_t = 0$$
$$v_{xx} + Li_{xt} + Ri_x = 0$$

We now multiply the first equation by L and subtract the second equation, getting

$$LCv_{tt} + LGv_t - v_{xx} - Ri_x = 0$$

and finally solving for v_{xx} (along with plugging in for $i_x = -Cv_t - Gv$ in the first equation) gives the desired result

$$v_{xx} = LCv_{tt} + LGv_t - Ri_x$$
$$= LCv_{tt} + LGv_t - R(-Cv_t - Gv)$$
$$= LCv_{tt} + (CR + LG)v_t + GRv$$

Lesson 16: The One Dimensional Wave Equation (Hyperbolic Eq)   109

2. From your knowledge of the various terms of the wave equation, what would you expect the solution of the following problem to look like at various values of time?

$$\text{PDE: } u_{tt} = u_{xx} - u_t \qquad 0 < x < 1, \ 0 < t < \infty$$

$$\text{BCs: } \begin{cases} u(0,t) = 0 \\ u(1,t) = 0 \end{cases} \qquad 0 < t < \infty$$

$$\text{ICs: } \begin{cases} u(x,0) = \sin(\pi x) \\ u_t(x,0) = 0 \end{cases} \qquad 0 < x < 1$$

Solution: The wave initially looks like the sine curve $\sin(\pi x)$ but vibrates to zero due to the friction (or damping) term u_t.

3. What would the solution of the following problem look like at various values of time? What is the physical interpretation of the problem?

$$\text{PDE: } u_{tt} = u_{xx} \qquad 0 < x < 1, \ 0 < t < \infty$$

$$\text{BCs: } \begin{cases} u(0,t) = 0 \\ u(1,t) = \sin(\pi t) \end{cases} \qquad 0 < t < \infty$$

$$\text{ICs: } \begin{cases} u(x,0) = 0 \\ u_t(x,0) = 0 \end{cases} \qquad 0 \le x \le 1$$

Solution: You might imagine this problem describing the motion of a vibrating string, which is initially at rest, but a person standing at the right end moving the string up and down like a sine wave. The motion is complicated but one can imagine waves moving down the string from right to left, sometimes bouncing off the left-hand end, which is fixed at zero, and then coming back only to collide with waves moving from the other direction. One can solve this problem analytically by either the Laplace transform (transforming either x or t actually) or by Duhamel's principle.

Lesson 16: The One Dimensional Wave Equation (Hyperbolic Eq)

> 4. How many solutions of $u_{tt} = u_{xx}$ can you find by looking for solutions of the form
> $$u(x,t) = e^{ax+bt} \ ?$$

Solution: Substituting
$$u(t,t) = e^{x+bt}$$
in the wave equation
$$u_{tt} = u_{xx}$$
one gets $b^2 = a^2$ or $b = \pm a$. Hence, we get solutions of the form

$$u(x,t) = e^{a(x-t)} \quad \text{and} \quad u(x,t) = e^{a(x+t)}$$

where a is any constant. The sum is also a constant and so we have more solutions
$$u(x,t) = c_1 e^{a(x-t)} + c_2 e^{a(x+t)}$$

Later we will see that the wave equation has (many) more solutions than this.

$$\Pi\Theta\Xi\Omega\Psi Z$$

Lesson 17 : The D'Alembert Solution of the Wave Equation

> 1. Verify that general D'Alembert solution
>
> $$u(x,t) = \frac{1}{2}\left[f(x-ct) + f(x+ct)\right] + \frac{1}{2c}\int_{x-ct}^{x+ct} g(\xi)d\xi$$
>
> satisfies the IVP
>
> PDE: $u_{tt} = c^2 u_{xx}$ $-\infty < x < \infty,\; 0 < t < \infty$
>
> ICs: $\begin{cases} u(x,0) = f(x) \\ u_t(x,0) = g(x) \end{cases}$ $-\infty < x < \infty$

Solution: The general formula for the derivative of an integral when the limits of integration are not constants, but functions is called the Leibniz formula given by

$$\frac{d}{dx}\int_{\alpha(x)}^{\beta(x)} g(\xi)d\xi = \frac{d\beta(x)}{dx} g[\beta(x)] - \frac{d\alpha(x)}{dx} g[\alpha(x)]$$

Remembering your multivariable calculus and how you compute partial derivatives, we see

$$u_t(x,t) = \frac{1}{2}\left[-cf'(x-ct) + cf'(x+ct)\right] + \frac{1}{2c}\left[c\,g(x+ct) - c\,g(x-ct)\right]$$

$$u_{tt}(x,t) = \frac{1}{2}\left[c^2 f''(x-ct) + c^2 f''(x+ct)\right] + \frac{1}{2}\left[c\,g'(x+ct) + c^2 g'(x-ct)\right]$$

$$u_x(x,t) = \frac{1}{2}\left[f'(x-ct) + f'(x+ct)\right] + \frac{1}{2c}\left[g(x+ct) + g(x-ct)\right]$$

$$u_{xx}(x,t) = \frac{1}{2}\left[f''(x-ct) + f''(x+ct)\right] + \frac{1}{2c}\left[g'(x+ct) + g'(x-ct)\right]$$

$$c^2 u_{xx}(x,t) = \frac{1}{2}\left[c^2 f''(x-ct) + c^2 f''(x+ct)\right] + \frac{1}{2}\left[c\,g'(x+ct) + c\,g'(x-ct)\right]$$

and upon close inspection, we see that $u_{tt} = c^2 u_{xx}$. The initial conditions can also be easily verified.

---

2. Substitute

$$\phi(x) = \frac{1}{2} f(x) - \frac{1}{2c} \int_{x_0}^{x} g(\xi) d\xi$$

$$\psi(x) = \frac{1}{2} f(x) + \frac{1}{2c} \int_{x_0}^{x} g(\xi) d\xi$$

in the general formula

$$u(x,t) = \phi(x-ct) + \psi(x+ct)$$

to get the D'Alembert solution.

Solution: By direct substitution, we get

$$u(x,t) = \phi(x-ct) + \psi(x+ct)$$

$$= \left[\frac{1}{2} f(x-ct) - \frac{1}{2c} \int_{x_0}^{x-ct} g(\xi) d\xi \right] + \left[\frac{1}{2} f(x+ct) + \frac{1}{2c} \int_{x_0}^{x+ct} g(\xi) d\xi \right]$$

$$= \frac{1}{2} \left[f(x-ct) + f(x+ct) \right] + \frac{1}{2c} \left[\int_{x_0}^{x+ct} g(\xi) d\xi - \int_{x_0}^{x-ct} g(\xi) d\xi \right]$$

$$= \frac{1}{2} \left[f(x-ct) + f(x+ct) \right] + \frac{1}{2c} \int_{x-ct}^{x+ct} g(\xi) d\xi$$

---

3. Solve the initial-value problem

PDE: $u_{tt} = u_{xx}$ $-\infty < x < \infty, \; 0 < t < \infty$

IC: $\begin{cases} u(x,0) = e^{-x^2} \\ u_t(x,0) = 0 \end{cases}$ $-\infty < x < \infty$

What does the solution look like for different values of time?

Lesson 17: The D'Alembert Solution of the Wave Equation

Solution: The D'Alembert solution gives

$$u(x,t) = \frac{1}{2}\left[f(x-ct) + f(x+ct)\right] + \frac{1}{2c}\int_{x-ct}^{x+ct} g(\xi)d\xi$$

$$= \frac{1}{2}\left[e^{-(x-t)^2} + e^{-(x+t)^2}\right]$$

---

4. Solve the initial-value problem

$$\text{PDE: } u_{tt} = u_{xx} \qquad -\infty < x < \infty,\ 0 < t < \infty$$

$$\text{IC: } \begin{cases} u(x,0) = 0 \\ u_t(x,0) = xe^{-x^2} \end{cases} \qquad -\infty < x < \infty$$

What does the solution look like for different values of time?

Solution: The D'Alembert solution gives

$$u(x,t) = \frac{1}{2}\left[f(x-ct) + f(x+ct)\right] + \frac{1}{2c}\int_{x-ct}^{x+ct} g(\xi)d\xi$$

$$= \frac{1}{2}\int_{x-t}^{x+t} \xi e^{-\xi^2} d\xi$$

$$= \frac{1}{2}\left[-\frac{1}{2}e^{-\xi^2}\Big|_{x-t}^{x+t}\right]$$

$$= \frac{1}{4}\left[e^{-(x-t)^2} - e^{-(x+t)^2}\right]$$

---

5. Algebraically solve for $\phi(x)$ and $\psi(x)$ from the equations

$$\phi(x) + \psi(x) = f(x)$$

$$-c\phi(x) + c\psi(x) = \int_{x_0}^{x} g(\xi)d\xi$$

to arrive at

$$\phi(x) = \frac{1}{2}f(x) - \frac{1}{2c}\int_{x_0}^{x} g(\xi)d\xi$$

$$\psi(x) = \frac{1}{2}f(x) + \frac{1}{2c}\int_{x_0}^{x} g(\xi)d\xi$$

Solution: Straightforward algebra.

ΣΠΘΩΛ

Lesson 18: More on the D'Alembert Solution

1. Solve the semi-infinite string problem

$$\text{PDE: } u_{tt} = c^2 u_{xx} \qquad 0 < x < \infty, \ 0 < t < \infty$$

$$\text{BC: } u(0,t) = 0 \qquad 0 < t < \infty$$

$$\text{ICs: } \begin{cases} u(x,0) = xe^{-x^2} \\ u_t(x,0) = 0 \end{cases} \qquad 0 < x < \infty$$

Solution: For the semi-infinite string we use the modified D-Alembert formula

$$u(x,t) = \begin{cases} \dfrac{1}{2}\left[f(x-ct) + f(x+ct)\right] + \dfrac{1}{2c}\int_{x-ct}^{x+ct} g(\xi)d\xi & ct \leq x \\ \dfrac{1}{2}\left[f(x+ct) - f(ct-x)\right] + \dfrac{1}{2c}\int_{ct-x}^{x+ct} g(\xi)d\xi & ct > x \end{cases}$$

$$= \begin{cases} \dfrac{1}{2}\left[(x-ct)e^{-(x-ct)^2} + (x+ct)e^{-(x+ct)^2}\right] & ct \leq x \\ \dfrac{1}{2}\left[(x+ct)e^{-(x+ct)^2} - (t-cx)e^{-(t-cx)^2}\right] & ct > x \end{cases}$$

---

2. The solution of the semi-infinite string problem in Problem 1 can also be written by

a) extending the IC to the entire x-axis by defining

$$u(x,0) = -xe^{-x^2} \quad -\infty < x < 0$$

$$u_t(x,0) = 0 \quad -\infty < x < 0$$

b) averaging the left and right moving waves.

c) taking the solution for $x \geq 0$

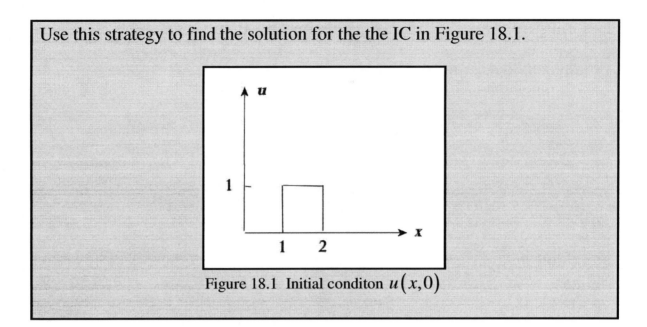

Use this strategy to find the solution for the the IC in Figure 18.1.

Figure 18.1 Initial conditon $u(x,0)$

Solution: The *left* moving wave from the initial wave

$$f(x,0) = \begin{cases} xe^{-x^2} & 0 < x < \infty \\ 0 & -\infty < x < 0 \end{cases}$$

is

$$u_-(x,t) = f(x+ct) = (x+ct)e^{-(x+ct)^2} \quad -\infty < x < \infty$$

and the *right* moving wave that results from the odd extension

$$f(x,0) = \begin{cases} -xe^{-x^2} & -\infty < x < 0 \\ 0 & 0 < x < \infty \end{cases}$$

is

$$u_+(x,t) = f(x-ct) = -(x-ct)e^{-(x-ct)^2}, \quad -\infty < x < \infty$$

Hence, we have

$$u(x,t) = \begin{cases} \dfrac{1}{2}\left[(x+ct)e^{-(x+ct)^2} + (x-ct)e^{-(x-ct)^2}\right] & ct \leq x \\ \dfrac{1}{2}\left[(x+ct)e^{-(x+ct)^2} - (t-cx)e^{-(t-cx)^2}\right] & ct > x \end{cases}$$

Lesson 18: More on the D'Alembert Solution

The values of x where $ct \leq x$ are those values of x where the wave has not yet bounced off the wall, while the values of x where $ct > x$ are those values of x where the wave has bounded off the wall.

Figure 18.2 shows the motion of an initial triangular wave with BCs $u(0,t) = 0$ by averaging the left and right moving waves of the function and its odd extension.

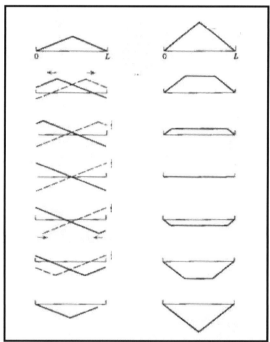

Figure 18.2 Averaging left and right moving waves.

---

3. Solve the semi-infinite string problem

$$\text{PDE: } u_{tt} = c^2 u_{xx} \qquad 0 < x < \infty, \ 0 < t < \infty$$

$$\text{BC: } u_x(0,t) = 0 \qquad 0 < t < \infty$$

$$\text{ICs: } \begin{cases} u(x,0) = f(x) \\ u_t(x,0) = 0 \end{cases} \qquad 0 < x < \infty$$

in a manner analogous of the way the semi-infinite string problem was solved in the lesson. What is the interpretation of the problem?

Solution We define the even extension of $f(x)$, $0 \leq x < \infty$ to the real line by

$$\tilde{f}(x) = \begin{cases} f(x) & x \geq 0 \\ f(-x) & x < 0 \end{cases}$$

We also extend $g(x) = 0$, $0 \leq x < \infty$ to the entire real line by $\tilde{g}(x) = 0$, $-\infty < x < \infty$. The solution $\tilde{u}$ of the extended problem

PDE: $\tilde{u}_{tt} = c^2 \tilde{u}_{xx}$ $\quad -\infty < x < \infty,\ 0 < t < \infty$

BC: $\tilde{u}_x(0,t) = 0$ $\quad 0 < t < \infty$

ICs: $\begin{cases} \tilde{u}(x,0) = f(x) \\ \tilde{u}_t(x,0) = 0 \end{cases}$ $\quad 0 < x < \infty$

is

$$\tilde{u}(x,t) = \frac{1}{2}\left[\tilde{f}(x-ct) + \tilde{f}(x+ct)\right]$$

We now define the function

$$u(x,t) = \tilde{u}(x,t),\ x \geq 0$$

and show that $u(x,t)$ satisfies the IBVP

PDE: $u_{tt} = c^2 u_{xx}$ $\quad 0 < x < \infty,\ 0 < t < \infty$

BC: $u_x(0,t) = 0$ $\quad 0 < t < \infty$

ICs: $\begin{cases} u(x,0) = f(x) \\ u_t(x,0) = 0 \end{cases}$ $\quad 0 < x < \infty$

The function u satisfies the PDE since $\tilde{u}$ does. The IC are also satisfied since

Lesson 18: More on the D'Alembert Solution 119

$$u(x,0) = \tilde{u}(x,0)$$
$$= \frac{1}{2}\left[\tilde{f}(x) + \tilde{f}(x)\right]$$
$$= \frac{1}{2}\left[f(x) + f(x)\right]$$
$$= f(x)$$

Finally, to show the BC $u_x(0,t) = 0$, we use the fact that if $\tilde{f}$ is an even extension then its derivative is an odd extension, i.e.

$$\tilde{f}'(-ct) = -\tilde{f}'(ct)$$

and by computing the derivative

$$\tilde{u}_x(x,t) = \frac{1}{2}\left[\tilde{f}'(x-ct) + \tilde{f}'(x+ct)\right]$$

we have

$$\tilde{u}_x(0,t) = \frac{1}{2}\left[\tilde{f}'(-ct) + \tilde{f}'(+ct)\right]$$
$$= \frac{1}{2}\left[-f'(ct) + f'(+ct)\right] = 0$$

and thus $u_x(0,t) = \tilde{u}_x(0,t) = 0$.

---

4. Suppose the vibration of a string is described by $u_{tt} = u_{xx}$ and has an initial displacement given by the diagram in Figure 18.3.

Lesson 18: More on the D'Alembert Solution

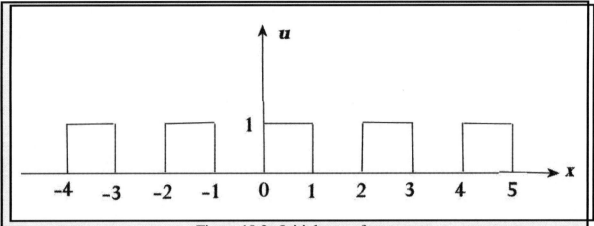

Figure 18.3 Initial wave form

Assuming the initial velocity of $u_t(x,0)=0$, describe the solution of this problem in the xt plane. Note that the IC is discontinuous.

Solution: Figure 18.4 shows the movement of the waves in the tx plane.

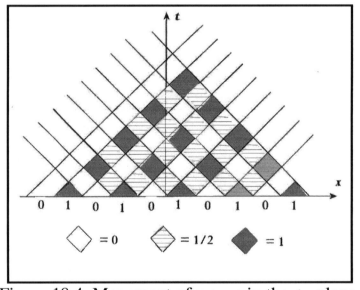

Figure 18.4 Movement of waves in the tx plane.

To visualize the solution imagine two initial half waves, one moving to the left the other to the right and adding the results. Try to visualize that happens to the wave on the interval $[0,1]$.

ΣΠΞΩZΨ

Lesson 19: Boundary Conditions Associated with the Wave Equation

1. From your intuition of the various kinds of BCs, draw a rough sketch of the solution to the following IBVP for various values of time.

$$\text{PDE:} \quad u_{tt} = c^2 u_{xx} \quad 0 < x < 1, \ 0 < t < \infty$$

$$\text{BC:} \quad u_x(0,t) = 0 \quad 0 \leq t < \infty$$

$$\text{ICs:} \quad \begin{cases} u(x,0) = f(x) \\ u_t(x,0) = 0 \end{cases} \quad 0 \leq x \leq 1$$

Solution: The boundary condition $u_x(0,t) = 0$ at $x = 0$ resembles waves moving to the left and hitting a wall at $x = 0$. The water moves in and down but the slope of the water at the wall is always zero. The waves then bounce back and move to the right. The reader can find good animations of this action online by 'googling' phrases like 'one-dimensional d'Alembert motion'.

2. From your intuition of the various kinds of BCs, draw a rough sketch of the solution to the following IBVP for various values of time.

$$\text{PDE:} \quad u_{tt} = u_{xx} \quad 0 < x < 1, \ 0 < t < \infty$$

$$\text{BC:} \quad \begin{cases} u(0,t) = 0 \\ u_x(1,t) = 0 \end{cases} \quad 0 \leq t < \infty$$

$$\text{ICs:} \quad \begin{cases} u(x,0) = \sin\left(\dfrac{\pi x}{2}\right) \\ u_t(x,0) = 0 \end{cases} \quad 0 \leq x \leq 1$$

Solution: We have seen that the solution of the wave equation $u_{tt} = u_{xx}$ when defined in 'free space', i.e. $-\infty < x < \infty$ can be interpreted as the average of two moving waves, one to the right the other to the left. However, when there are boundaries like in this problem, the waves hit the walls at $x = 0$ and $x = 1$ and bounce back in some manner, exactly how they react to the wall depends on the boundary conditions ($u(0,t) = 0$ is like a rope tide down at the end, whereas $u_x(0,t)=0$ is the way water behaves as it reaches the wall). But waves bouncing back and forth become difficult, hitting each other as they run into each other, and so a better way to analyze wave problems on finite intervals, like $0<x<1$ is in terms of stand waves.

In this problem the initial wave is zero at the left $(x=0)$ and 1 at the right $(x=1)$. The shape of solutions of this problem are the eigenfunctions of the Sturm Liouville problem

$$X'' + \lambda^2 X = 0 \quad 0<x<1$$
$$X(0)=0$$
$$X'(1)=0$$

which are $\cos(n\pi/2)$ and since our IC is the first eigenfunction, we suspect a solution to be a simple vibration like

$$u(x,t) = \cos\left(\frac{\pi t}{2}\right)\sin\left(\frac{\pi x}{2}\right)$$

You can verify that this function satisfies the IBVP.

3. What is the general nature of the BC

$$u_x(0,t) + hu(0,t) = 0$$

when h is small, when h is large.

Solution: In conjunction with the wave equation, when $h=0$, we have that $u_x(0,t)=0$. Have you ever seen waves lapping up against a wall? Provided the waves don't splash, the slope of the waves against the wall is zero, hence simulating the BC $u_x(0,t)=0$. When h is large, the solution at $x=0$ pays more attention to whether $u(1,t)=0$.

Lesson 19: Boundary Conditions with the Wave Equation

4. From your intuition of the various kinds of BCs, draw a rough sketch of the solution to the following IBVP for various values of time.

$$\text{PDE: } u_{tt} = u_{xx} \qquad 0 < x < 1, \ 0 < t < \infty$$

$$\text{BC: } \begin{cases} u(0,t) = 0 \\ u_x(1,t) = 0 \end{cases} \qquad 0 < t < \infty$$

$$\text{ICs: } \begin{cases} u(x,0) = x \\ u_t(x,0) = 0 \end{cases} \qquad 0 \le x \le 1$$

Solution: The problem describe a sting whose initial shape is $u(x,0) = x$ that initially have no motion $(u_t(x,0) = 0)$. The left end is fixed at 0 but the right end is free to move up and down but has slope 0 at the right. We know that the PDE and BC give rise to the Sturm Liouville problem

$$X'' + \lambda^2 X = 0 \quad 0 < x < 1$$
$$X(0) = 0$$
$$X'(1) = 0$$

that has eigenfunctions

$$X_n(x) = \sin\left(\frac{n\pi}{2}\right), \quad n = 1, 2, \ldots$$

These are the functions that give rise to simple harmonic motion of the type

$$u_n(x,t) = a_n T_n(t) X_n(x) = a_n \sin\left(\frac{n\pi x}{2}\right) \cos\left(\frac{n\pi x}{2}\right)$$

Hence, to solve this problem, we simply expand the IC

$$x = a_1 \sin\left(\frac{\pi x}{2}\right) + a_2 \sin\left(\frac{2\pi x}{2}\right) + a_3 \sin\left(\frac{3\pi x}{2}\right) + \cdots \quad 0 \le x \le 1$$

where

$$a_n = \frac{2}{\pi}\int_0^1 x\sin\left(\frac{n\pi x}{2}\right)dx$$

and then the solution is

$$u(x,t) = \sum_{n=1}^{\infty} a_n \sin\left(\frac{n\pi x}{2}\right)\cos\left(\frac{n\pi x}{2}\right)$$

$$\Sigma \mathrm{P} \Xi \Psi \mathrm{Z} \Omega$$

Lesson 20: The Finite Vibrating String (Standing Waves)

> 1. Find the solution to the following IBVP. Is the solution periodic in time, if so what is the period?
>
> $$\text{PDE: } u_{tt} = \alpha^2 u_{xx} \qquad 0 < x < L,\ 0 < t < \infty$$
>
> $$\text{BC: } \begin{cases} u(0,t) = 0 \\ u(L,t) = 0 \end{cases} \qquad 0 < t < \infty$$
>
> $$\text{ICs: } \begin{cases} u(x,0) = \sin\left(\dfrac{\pi x}{L}\right) + 0.5\sin\left(\dfrac{3\pi x}{L}\right) \\ u_t(x,0) = 0 \end{cases} \qquad 0 \le x \le L$$

Solution: The PDE and BCs of this problem give eigenfunctions

$$X_n(x) = \sin\left(\frac{n\pi x}{L}\right),\ n = 1, 2, \ldots$$

and so the solution has the form

$$u(x,t) = \sum_{n=1}^{\infty} \sin\left(\frac{n\pi x}{L}\right)\left[a_n \sin\left(\frac{n\pi \alpha t}{L}\right) + b_n \cos\left(\frac{n\pi \alpha t}{L}\right)\right]$$

Since the initial velocity $u_t(x,0) = 0$, we have that all the coefficients $a_n^s = 0$ and that the b_n^s are the Fourier coefficients of the of IC $u(x,0)$ or .

$$b_n = \begin{cases} 1 & n = 1 \\ 1/2 & n = 3 \\ 0 & \text{all other } n \end{cases}$$

Hence, the solution is

$$u(x,t) = \sum_{n=1}^{\infty} \sin\left(\frac{n\pi x}{L}\right)\left[a_n \sin\left(\frac{n\pi\alpha t}{L}\right) + b_n \cos\left(\frac{n\pi\alpha t}{L}\right)\right]$$

$$= \sin\left(\frac{\pi x}{L}\right)\cos\left(\frac{\pi\alpha t}{L}\right) + \frac{1}{2}\sin\left(\frac{3\pi x}{L}\right)\cos\left(\frac{3\pi\alpha t}{L}\right)$$

---

2. Find the solution to the following IBVP. Is the solution periodic in time, if so what is the period?

$$\text{PDE: } u_{tt} = \alpha^2 u_{xx} \quad 0 < x < L, \ 0 < t < \infty$$

$$\text{BC: } \begin{cases} u(0,t) = 0 \\ u(L,t) = 0 \end{cases} \quad 0 < t < \infty$$

$$\text{ICs: } \begin{cases} u(x,0) = 0 \\ u_t(x,0) = \sin(3\pi x / L) \end{cases} \quad 0 \le x \le L$$

Solution: The PDE and BCs of this problem give eigenfunctions

$$X_n(x) = \sin\left(\frac{n\pi x}{L}\right), \ n = 1, 2, \ldots$$

and so the solution has the form

$$u(x,t) = \sum_{n=1}^{\infty} \sin\left(\frac{n\pi x}{L}\right)\left[a_n \sin\left(\frac{n\pi\alpha t}{L}\right) + b_n \cos\left(\frac{n\pi\alpha t}{L}\right)\right]$$

where the a_n^s are the the Fourier coefficients of the initial velocity $u_t(x,0)$ and the b_n^s are the Fourier coefficients of the IC $u(x,0)$. Hence we have all the $b_n^s = 0$ and

$$a_n = \begin{cases} \dfrac{L}{3\pi\alpha} & n = 3 \\ 0 & n \ne 3 \end{cases}$$

Hence, the solution is

Lesson 20: The Finite Vibrating String (Standing Waves)

$$u(x,t) = \left(\frac{L}{3\pi\alpha}\right)\sin\left(\frac{3\pi x}{L}\right)\sin\left(\frac{3\pi\alpha t}{L}\right)$$

---

3. Show that for $\lambda \geq 0$ in Figure 20.2 in the book, the solutions $X(x)T(t)$ are either unbounded or zero.

Solution: When $\lambda = 0$ both functions $X(x)$ and $T(t)$ are linear functions in x and t respectively. The function $T(t)$ must increase with respect to t and so goes to infinity, which means the product $X(x)T(t)$ does not remain bounded. When $\lambda > 0$ one of the terms of both $X(x)$ and $T(t)$ is an exponential growing exponent in x and t, respectively and so the product $X(x)T(t)$ does not remain bounded.

---

4. What is the solution of the IBVP

PDE: $u_{tt} = \alpha^2 u_{xx}$ $0 < x < L,\ 0 < t < \infty$

BC: $\begin{cases} u(0,t) = 0 \\ u(L,t) = 0 \end{cases}$ $0 < t < \infty$

ICs: $\begin{cases} u(x,0) = \sin(3\pi x/L) \\ u_t(x,0) = (3\pi\alpha/L)\sin(3\pi x/L) \end{cases}$ $0 \leq x \leq L$

Solution: With the given PDE and BCs the solution has the general form

$$u(x,t) = \sum_{n=1}^{\infty} \sin\left(\frac{n\pi x}{L}\right)\left[a_n \sin\left(\frac{n\pi\alpha t}{L}\right) + b_n \cos\left(\frac{n\pi\alpha t}{L}\right)\right]$$

where the coefficients a_n and b_n are the Fourier coefficients of the ICs. In this problem, we have

$$a_n = \begin{cases} 1 & n = 3 \\ 0 & n \neq 3 \end{cases}$$

$$b_n = \begin{cases} 1 & n = 3 \\ 0 & n \neq 3 \end{cases}$$

and so the solution is

$$u(x,t) = \sin\left(\frac{3\pi x}{L}\right)\left[\sin\left(\frac{3\pi\alpha t}{L}\right) + \cos\left(\frac{3\pi\alpha t}{L}\right)\right]$$

The main observation to be made here is that the string vibrates with simple harmonic motion with circular frequency $3\pi\alpha/L$ oscillations per 2π units of time, i.e. everywhere along the string vibrates with the same frequency and the vibrating string maintains is original 'profile'.

---

5. A guitar string of length is pulled upward at the middle so that it reaches a height of h. Assuming the position of the string is initially

$$u(x,0) = \begin{cases} 2hx & 0 \le x < 0.5 \\ 2h(1-x) & 0.5 \le x \le 1 \end{cases}$$

as drawn in Figure 20.1, what is the subsequent motion of the string after it is released?

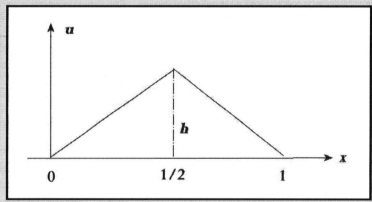

Figure 20.1 Initial position of the guitar string

Solution: The solution has the form

$$u(x,t) = \sum_{n=1}^{\infty} \sin(n\pi x)\left[a_n \sin(n\pi\alpha t) + b_n \cos(n\pi\alpha t)\right]$$

Plugging this into the ICs gives

Lesson 20: The Finite Vibrating String (Standing Waves)

$$u(x,0) = \sum_{n=1}^{\infty} b_n \sin(n\pi x) = \begin{cases} 2hx & 0 \le x < 0.5 \\ 2h(1-x) & 0.5 \le x \le 1 \end{cases}$$

$$u_t(x,0) = \sum_{n=1}^{\infty} a_n (n\pi\alpha) \sin(n\pi x) = 0$$

from which we (by inspection) get $a_n = 0$. To find the coefficients b_n we multiply the first above equation by

$$\sin(m\pi x)$$

and integrate each side of the equation with respect to x from 0 to 1, getting

$$b_m \int_0^1 \sin^2(m\pi x)\, dx = 2h\int_0^{1/2} x\sin(m\pi x)\, dx + 2h\int_{1/2}^1 (1-x)\sin(m\pi x)\, dx$$

Solving for b_m and with the help of wolframalpha, we find

$$b_m = \frac{2h\left[\int_0^{1/2} x\sin(m\pi x)\, dx + \int_{1/2}^1 (1-x)\sin(m\pi x)\, dx\right]}{\int_0^1 \sin^2(m\pi x)\, dx} = \begin{cases} 8h/(n\pi)^2 & m=1,5,9,\ldots \\ -8h/(n\pi)^2 & m=3,7,11 \\ 0 & m=0,2,4,6, \end{cases}$$

Hence, the solution is

$$u(x,t) = \left(\frac{8h}{\pi^2}\right)\left[\sin(\pi x)\cos(\pi t) - \frac{1}{(3)^2}\sin(3\pi x)\cos(3\pi t) + \cdots\right]$$

---

6. Solve the damped vibrating string problem

$$\text{PDE: } u_{tt} = \alpha^2 u_{xx} - \beta u \quad 0 < x < 1,\ 0 < t < \infty$$

BCs: $\begin{cases} u(0,t) = 0 \\ u(1,t) = 0 \end{cases} \quad 0 < t < \infty$

ICs: $\begin{cases} u(x,0) = f(x) \\ u_t(x,0) = 0 \end{cases} \quad 0 \le x \le L$

Solution: Looking for standing wave solutions

$$u(x,t) = X(x)T(t)$$

the PDE becomes

$$XT'' = \alpha^2 X''T - \beta XT$$

or

$$\frac{T''}{\alpha^2 T} = \frac{X''}{X} - \frac{\beta}{\alpha^2}$$

Since the left hand side of this equation depends only on t and the right hand since only on x, then in order they are equal for all x and t, they both must be equal to a constant, so setting both the left and right hand sides of this equation to a constant we call $-k^2$, we have

$$T'' + (\alpha k)^2 T = 0$$

$$X'' + \left(k^2 - \frac{\beta}{\alpha^2}\right) X = 0$$

Calling

$$\lambda^2 = k^2 - \frac{\beta}{\alpha^2}$$

we now have that $X(x)$ must satisfy the BVP

ODE: $X'' + \lambda^2 X = 0$

BCs: $\begin{cases} X(0) = 0 \\ X(1) = 0 \end{cases}$

which has eigenvalues $\lambda_n = n\pi$, $n = 1, 2,...$ and eigenfunctions $X_n(x) = \sin(n\pi x)$. We now find the constant k from the equation

$$n^2\pi^2 = k^2 - \frac{\beta}{\alpha^2} \Rightarrow k_n = n^2\pi^2 + \frac{\beta}{\alpha^2}$$

and so the equation

$$T'' + (\alpha k)^2 T = 0$$

becomes

Lesson 20: The Finite Vibrating String (Standing Waves)

$$T_n'' + \left[(n\pi\alpha)^2 + \beta\right]T_n = 0$$

whose general solutions is

$$T_n(t) = c_1 \cos(\gamma t) + c_2 \sin(\gamma t)$$

where c_1, c_2 are arbitrary constants and $\gamma_n = (n\pi\alpha)^2 + \beta$, where we assume $\beta > 0$ is sufficiently small so that all the $\gamma_n > 0$ for all $n = 1, 2, \ldots$. The solution of the IBVP then has the form

$$u(x,t) = \sum_{n=1}^{\infty} b_n \sin(n\pi x) \cos\left[\sqrt{(n\pi\alpha)^2 + \beta}\; t\right]$$

where the coefficients b_n are the Fourier coefficients of the IC

$$u(x,0) = f(x)$$

The reader can verify that this function is the solution of the IBVP. Note how the extra forcing term $-\beta u$ on the string (kind of like a Hooke's law trying to restore the sting to zero) makes the vibrations of the string faster; i.e. compare the faster vibration

$$\cos\left[\left((n\pi\alpha)^2 + \beta\right)t\right]$$

versus the slower vibration

$$\cos\left[(n\pi\alpha)^2 t\right]$$

---

7. How would you solve the nonhomogeneous PDE with given boundary and initial conditions?

PDE: $u_{tt} = \alpha^2 u_{xx} + Kx \quad 0 < x < 1, \; 0 < t < \infty$

BCs: $\begin{cases} u(0,t) = 0 \\ u(1,t) = 0 \end{cases} \quad 0 < t < \infty$

ICs: $\begin{cases} u(x,0) = f(x) \\ u_t(x,0) = 0 \end{cases} \quad 0 \leq x \leq L$

Lesson 20: The Finite Vibrating String (Standing Waves)

Solution: We seek a solution in the form

$$u(x,t) = \sum_{n=1}^{\infty} T_n(t) X_n(x) = \sum_{n=1}^{\infty} T_n(t) \sin(n\pi x)$$

where the

$$X_n(x) = \sin(n\pi x)$$

are the eigenfunctions of the BVP

$$\text{ODE: } X'' + \lambda^2 X = 0$$

$$\text{BCs: } \begin{cases} X(0) = 0 \\ X(1) = 0 \end{cases}$$

To find the functions of time $T_n(t)$ we find the eigenfunction expansion of the nonhomogeneous term

$$Kx = \sum_{n=1}^{\infty} f_n(t) \sin(n\pi x)$$

getting Fourier coefficients

$$f_n(t) = \frac{\int_0^1 Kx \sin(n\pi x)\, dx}{\int_0^1 \sin^2(n\pi x)\, dx} = \frac{(-1)^{n+1} 2K}{n\pi}$$

The first few of these (constant) functions are

$$(f_1(t), f_2(t), f_3(t), f_4(t)\ldots) = \left(\frac{2K}{\pi}, \frac{-2K}{2\pi}, \frac{2K}{3\pi}, \frac{-2K}{4\pi}, \ldots\right)$$

We can now find the functions $T_n(t)$ by solving

$$\text{ODE: } T_n'' + (n\pi\alpha)^2 T_n = f_n(t)$$

$$\text{IC: } \begin{cases} T_n(0) = a_n \\ T_n'(0) = 0 \end{cases}$$

where the constants a_n in the ICs are the Fourier coefficients in the series

Lesson 20: The Finite Vibrating String (Standing Waves)

$$f(x) = \sum_{n=1}^{\infty} a_n \sin(n\pi x)$$

which are found from

$$a_n = \frac{\int_0^1 f(x)\sin(n\pi x)\,dx}{\int_0^1 \sin^2(n\pi x)\,dx} = \frac{1}{2}\int_0^1 f(x)\sin(n\pi x)\,dx$$

We then solve for $T_n(t)$ and plug these functions in

$$u(x,t) = \sum_{n=1}^{\infty} T_n(t) X_n(x) = \sum_{n=1}^{\infty} T_n(t)\sin(n\pi x)$$

$$\text{ΟΙΠΩΨΞ}$$

Lesson 21: The Vibrating Beam (Fourth-Order PDE)

1. Solve the following cantilever-beam problem.

$$\text{PDE: } u_{tt} + u_{xxxx} = 0 \quad 0 < x < 1, \ 0 < t < \infty$$

$$\text{BCs: } \begin{cases} u(0,t) = 0 \\ u_x(0,t) = 0 \\ u_{xx}(1,t) = 0 \\ u_{xxx}(1,t) = 0 \end{cases} \quad 0 < t < \infty$$

$$\text{ICs: } \begin{cases} u(x,0) = f(x) \\ u_t(x,0) = g(x) \end{cases} \quad 0 \leq x \leq 1$$

Hint: Although the eigenfunctions $X_n(x)$ are not the usual functions (like sines and cosines) as they were in the case of the heat or wave equations, we can still separate variables, arriving at a Sturm-Liouville eigenvalue problem that gives rise to the solution as a sum of eigenfunctions times functions of time.

Solution: Looking for solutions of the form $u(x,t) = X(x)T(t)$, we arrive at

$$XT'' + X''''T = 0$$

and dividing by XT gives

$$\frac{T''}{T} + \frac{X''''}{X} = 0$$

or

$$\frac{T''}{T} = -\frac{X''''}{X}$$

and setting both sides of the above equation to a constant, which we judiciously pick as $-\lambda^2$, we arrive at the two ODEs

Lesson 21: The Vibrating Beam (Fourth-Order PDE)

$$\frac{d^2 T}{dt^2} + \lambda^2 T = 0$$

$$\frac{d^4 X}{dx^4} - \lambda^2 X = 0$$

We choose the notation for the separation constant $-\lambda^2$ since it must be less than or equal to zero, else $T(t)$ is not periodic. The two ODEs have general solutions

$$T(t) = A\sin(\lambda t) + B\cos(\lambda t)$$
$$X(x) = C\sin(\sqrt{\lambda}x) + D\cos(\sqrt{\lambda}x) + E\sinh(\sqrt{\lambda}x) + F\cosh(\sqrt{\lambda}x)$$

To determine λ, we plug the general solution $X(t)$ in the BCs getting

$$X(0) = D + F = 0$$
$$X'(0) = C\sqrt{\lambda} + E\sqrt{\lambda} = 0$$
$$X''(1) = -C\lambda\sin(\sqrt{\lambda}) - D\lambda\cos(\sqrt{\lambda}) + E\lambda\sinh(\sqrt{\lambda}) + F\lambda\cosh(\sqrt{\lambda}) = 0$$
$$X'''(1) = -C\lambda^{3/2}\sin(\sqrt{\lambda}) + D\lambda^{3/2}\cos(\sqrt{\lambda}) + E\lambda^{3/2}\sinh(\sqrt{\lambda}) + F\lambda^{3/2}\cosh(\sqrt{\lambda}) = 0$$

The first two of these equations give us

$$D = -F, \quad C = -E$$

and plugging these values into the last two equation gives two equations, which we write in matrix form

$$\begin{bmatrix} \sin(\sqrt{\lambda}) + \sinh(\sqrt{\lambda}) & \cos(\sqrt{\lambda}) + \cosh(\sqrt{\lambda}) \\ \sqrt{\lambda}\sin(\sqrt{\lambda}) + \sqrt{\lambda}\sinh(\sqrt{\lambda}) & -\sqrt{\lambda}\cos(\sqrt{\lambda}) + \sqrt{\lambda}\cosh(\sqrt{\lambda}) \end{bmatrix} \begin{bmatrix} E \\ F \end{bmatrix} = \begin{bmatrix} 0 \\ 0 \end{bmatrix}$$

From these two simultaneous equations, we must have that the determinant of the coefficient matrix is zero, else $E = F = 0$ which would imply all the four values $C = D = E = F = 0$, giving the unaccepted zero solution $X(x) = 0$. So, setting the determinant to zero; i.e.

$$\begin{vmatrix} \sin(\sqrt{\lambda}) + \sinh(\sqrt{\lambda}) & \cos(\sqrt{\lambda}) + \cosh(\sqrt{\lambda}) \\ \sqrt{\lambda}\sin(\sqrt{\lambda}) + \sqrt{\lambda}\sinh(\sqrt{\lambda}) & -\sqrt{\lambda}\cos(\sqrt{\lambda}) + \sqrt{\lambda}\cosh(\sqrt{\lambda}) \end{vmatrix} = 0$$

we arrive at a (horrible) equation for which we can solve (using a computer of course) for the eigenvalues $\sqrt{\lambda_n}$, $n = 1, 2, \ldots$, which turn out to be a sequence of positive real numbers converging to $+\infty$. After finding the eigenvalues $\sqrt{\lambda_n}$, $n = 1, 2, \ldots$ and after a little algebra, we find

$$E_n = \cos\sqrt{\lambda_n} + \cosh\sqrt{\lambda_n}$$
$$F_n = \sin\sqrt{\lambda_n} + \sinh\sqrt{\lambda_n}$$

and thus we can find $D_n = -F_n$, $C_n = -E_n$. Knowing C_n, D_n, E_n, F_n (a certain number of them anyway using a computer) we can find the eigenfunctions

$$X_n(x) = C_n \sin(\sqrt{\lambda_n}x) + D_n \cos(\sqrt{\lambda_n}x) + E_n \sinh(\sqrt{\lambda_n}x) + F_n \cosh(\sqrt{\lambda_n}x)$$

Using λ_n we now solve the ODE in T to get

$$T_n(t) = a_n \cos(\lambda_n t) + b_n \sin(\lambda_n t)$$

and so the fundamental solutions of the problem are

$$u(x,t) = \sum_{n=1}^{\infty} u_n(x,t)$$
$$= \sum_{n=1}^{\infty} X_n(x) T_n(t)$$
$$= \sum_{n=1}^{\infty} \left[C_n \sin(\sqrt{\lambda_n}x) + D_n \cos(\sqrt{\lambda_n}x) + E_n \sinh(\sqrt{\lambda_n}x) + F_n \cosh(\sqrt{\lambda_n}x) \right]$$
$$\times \left[a_n \cos(\lambda_n t) + b_n \sin(\lambda_n t) \right]$$

The final job is to find the coefficients a_n and b_n so that $u(x,t)$ satisfies the IC. Plugging $u(x,0) = f(x), u_t(x,0) = g(t)$ into the IC gives

Lesson 21: The Vibrating Beam (Fourth-Order PDE)

$$u(x,0) = \sum_{n=1}^{\infty} b_n X_n(x) = f(x)$$

$$u_t(x,0) = \sum_{n=1}^{\infty} \lambda_n a_n X_n(x) = g(x)$$

We now use the fact that the functions $X_n(x)$ are orthogonal (the Sturm Liouville theory of BVPs for ODEs extends to 4th order equations as well as 2nd order). In other words

$$\int_0^1 X_m(x) X_n(x) dx = \begin{cases} 0 & m \neq n \\ \neq 0 & m = n \end{cases}$$

and so

$$a_n = \frac{\frac{1}{\lambda_n}\int_0^1 g(x) X_n(x) dx}{\int_0^1 X_n^2(x) dx}$$

$$b_n = \frac{\int_0^1 f(x) X_n(x) dx}{\int_0^1 X_n^2(x) dx}$$

---

2. What is the solution to the simply supported beam (at both ends) with ICs

ICs: $\begin{cases} u(x,0) = \sin(\pi x) \\ u_t(x,0) = \sin(\pi x) \end{cases} \quad 0 \leq x \leq 1$

Solution: The idea is to determine the coefficients C, D, E, F and frequencies $\sqrt{\lambda}$ in the general form

$$X(x) = C \sin(\sqrt{\lambda} x) + D \cos(\sqrt{\lambda} x) + E \sinh(\sqrt{\lambda} x) + F \cosh(\sqrt{\lambda} x)$$

that will satisfy the four BCs for a simply supported bean; that is

$$X(0) = 0$$
$$X''(0) = 0$$
$$X(1) = 0$$
$$X''(1) = 0$$

After a brief examination we see the only function having all four of these properties is

$$X_n(x) = \sin(n\pi x), \quad n = 1, 2, \ldots$$

Hence the solution has the general form

$$u(x,t) = \sum_{n=1}^{\infty} \sin(n\pi x) \left[a_n \sin\left[(n\pi)^2 t\right] + b_n \cos\left[(n\pi)^2 t\right] \right]$$

where we are left to determine the coefficients a_n, b_n. So, plugging this expression in the ICs gives us

$$u(x,0) = \sum_{n=1}^{\infty} b_n \sin(n\pi x) = \sin(\pi x)$$

$$u_t(x,0) = \sum_{n=1}^{\infty} \pi^2 a_n \sin(n\pi x) = \sin(\pi x)$$

which after a quick examination, we see $b_1 = 1$, $a_1 = 1/\pi^2$, all other coefficients zero. Hence the vibrations of this beam are

$$u(x,t) = \sin(\pi x) \left[\cos(\pi^2 t) + \frac{1}{\pi^2} \sin(\pi^2 t) \right]$$

which says the beam has a profile of $\sin(\pi x)$ and a vibrates with a frequency of $\pi^2/2\pi = \pi/2$ oscillations per seconds and period $2/\pi$ seconds.

---

3. What is the solution of a simply supported beam with ICs

$$\text{ICs:} \quad \begin{cases} u(x,0) = 1 - x^2 \\ u_t(x,0) = 0 \end{cases} \quad 0 \leq x \leq 1$$

Solution: In Problem 2 we found the possible shapes of a simply supported beam to be

Lesson 21: The Vibrating Beam (Fourth-Order PDE)

$$X(x) = \sin(n\pi x), \quad n=1,2,...$$

(we don't allow $n=0$ since that gives $X(x) \equiv 0$) and so the vibrations have the form

$$u(x,t) = \sum_{n=1}^{\infty} \sin(n\pi x)\left[a_n \cos\left[(n\pi)^2 t\right] + b_n \sin\left[(n\pi)^2 t\right]\right]$$

where the coefficients a_n, b_n are determined from the ICs. Plugging u in the ICs gives

ICs: $\begin{cases} u(x,0) = \sum_{n=1}^{\infty} a_n \sin(n\pi x) = 1 - x^2 \\ u_t(x,0) = \sum_{n=1}^{\infty} b_n \pi^2 \sin(n\pi x) = 0 \end{cases}$ $\quad 0 \le x \le 1$

which yields $b_n = 0$ for all $n = 1,2,...$ and a_n are found by multiply each side of the respective equation by $\sin(m\pi x)$ and integrating each side of the equation from 0 to 1, giving

$$\int_0^1 (1-x^2)\sin(m\pi x)\,dx = a_m \int_0^1 \sin^2(m\pi x)\,dx$$

or

$$a_m = \frac{\int_0^1 (1-x^2)\sin(m\pi x)\,dx}{\int_0^1 \sin^2(m\pi x)\,dx} = \begin{cases} \dfrac{2(m^2\pi^2 + 4)}{m^3 \pi^3} & m=1,3,... \\ \dfrac{2}{m\pi} & m=2,4,... \end{cases}$$

Hence, the solution

$$u(x,t) = \sum_{n=1}^{\infty} a_n \cos\left[(n\pi)^2 t\right] \sin(n\pi x)$$

The first term in the series is

$$\text{first term} = 0.89 \cos(\pi^2 t) \sin(\pi x)$$

which an approximation of the vibration of the beam. Note that this first term gives us an approximation of frequency $(2/\pi \text{ hertz})$ and period $(2/\pi \text{ second})$.

> **4.** Let the left end $(x=0)$ of a beam be rigidly fastened to a wall and let the right end $(x=1)$ be simply fastened according to the BCs shown in Figure 21.3 in the text. Solve the beam problem with these BCs and tell how to find the natural frequencies of vibration of the beam. Knowing the natural frequencies of the beam is important to know since various kinds of inputs of the same frequency give rise to resonance.

Solution: The natural frequencies are the eigenvalues of the Sturm-Liouville problem

$$\text{ODE:} \quad \frac{d^4 X}{dx^4} - \lambda^2 X = 0$$

$$\text{BCs:} \quad \begin{cases} X(0) = 0 \\ X'(0) = 0 \\ X(1) = 0 \\ X''(1) = 0 \end{cases}$$

The general solution of the differential equation is

$$X(x) = C\sin(\sqrt{\lambda}x) + D\cos(\sqrt{\lambda}x) + E\sinh(\sqrt{\lambda}x) + F\cosh(\sqrt{\lambda}x)$$

Plugging this function in the BCs gives

$$X(0) = D + F = 0$$
$$X'(0) = C\sqrt{\lambda} + E\sqrt{\lambda} = 0$$
$$X(1) = C\sin(\sqrt{\lambda}) + D\cos(\sqrt{\lambda}) + E\sinh(\sqrt{\lambda}) + F\cosh(\sqrt{\lambda}) = 0$$
$$X''(1) = -C\lambda\sin(\sqrt{\lambda}) - D\lambda\cos(\sqrt{\lambda}) + E\lambda\sinh(\sqrt{\lambda}) + F\lambda\cosh(\sqrt{\lambda}) = 0$$

The first two equations give $D = -E, C = -F$. Plugging these values in the last two equations gives

Lesson 21: The Vibrating Beam (Fourth-Order PDE)

$$\left[-\sin\left(\sqrt{\lambda}\right)+\cosh\left(\sqrt{\lambda}\right)\right]F+\left[-\cos\left(\sqrt{\lambda}\right)+\sinh\left(\sqrt{\lambda}\right)\right]E=0$$

$$\left[\lambda\sin\left(\sqrt{\lambda}\right)+\lambda\cosh\left(\sqrt{\lambda}\right)\right]F+\left[\lambda\cos\left(\sqrt{\lambda}\right)+\lambda\sinh\left(\sqrt{\lambda}\right)\right]E=0$$

There will be nonzero solutions for A, B, C, D unless the values of λ are chosen so that the determinant of the coefficient matrix is zero, i.e.

$$\begin{vmatrix} -\sin\left(\sqrt{\lambda}\right)+\cosh\left(\sqrt{\lambda}\right) & -\cos\left(\sqrt{\lambda}\right)+\sinh\left(\sqrt{\lambda}\right) \\ \lambda\sin\left(\sqrt{\lambda}\right)+\lambda\cosh\left(\sqrt{\lambda}\right) & \lambda\cos\left(\sqrt{\lambda}\right)+\lambda\sinh\left(\sqrt{\lambda}\right) \end{vmatrix} = 0$$

These equations can be solve on a computer getting eigenvalues $\sqrt{\lambda_n}$, $n=1,2,...$ which represent the frequency of vibrations of the bean. We can with a little algebraic work solve the above equations for E, F getting

$$E_n = \cos\left(\sqrt{\lambda_n}\right) + \cosh\left(\sqrt{\lambda_n}\right)$$

$$F_n = -\sin\left(\sqrt{\lambda_n}\right) - \sinh\left(\sqrt{\lambda_n}\right)$$

and hence $D_n = -E_n, C_n = -F_n$. We now have the solution in the form

$$u(x,t) = \sum_{n=1}^{\infty} u_n(x,t)$$

$$= \sum_{n=1}^{\infty} X_n(x) T_n(t)$$

$$= \sum_{n=1}^{\infty} \left\{\left[-\cos\left(\sqrt{\lambda}x\right)+\sinh\left(\sqrt{\lambda}x\right)\right]E_n + \left[-\sin\left(\sqrt{\lambda}x\right)+\cosh\left(\sqrt{\lambda}x\right)\right]F_n\right\} \left[a_n \cos(\lambda_n t) + b_n \sin(\lambda_n t)\right]$$

We now substitute this series in the IC to find the a_n, b as we did in Problem 1.

$$\Omega\Sigma\Xi\Psi Z\Upsilon$$

Lesson 22: Dimensionless Problems

> 1. Transform the vibrating string problem
>
> $$\text{PDE: } u_t = \alpha^2 u_{xx} \qquad 0 < x < L, \ 0 < t < \infty$$
>
> $$\text{BC: } \begin{cases} u(0,t) = 0 \\ u(L,t) = 0 \end{cases} \qquad 0 < t < \infty$$
>
> $$\text{IC: } \begin{cases} u(x,0) = \sin\left(\dfrac{\pi x}{L}\right) + \dfrac{1}{2}\sin\left(\dfrac{3\pi x}{L}\right) \\ u_t(x,0) = 0 \end{cases} \qquad 0 \leq x \leq L$$
>
> with independent variables x, t to dimensionless form
>
> $$\text{PDE: } u_{\tau\tau} = u_{\xi\xi} \qquad 0 < \xi < 1, \ 0 < \tau < \infty$$
>
> $$\text{BC: } \begin{cases} u(0,\tau) = 0 \\ u(1,\tau) = 0 \end{cases} \qquad 0 < \tau < \infty$$
>
> $$\text{IC: } \begin{cases} u(\xi,0) = \sin(\pi\xi) + \dfrac{1}{2}\sin(3\pi\xi) \\ u_\tau(\xi,0) = 0 \end{cases} \qquad 0 \leq \xi \leq 1$$
>
> in dimensionless variables ξ, τ.

Solution: Making the change of variables

$$\tau = \left(\frac{\alpha}{L}\right)t, \ \xi = \frac{x}{L} \quad \text{or} \quad t = \left(\frac{L}{\alpha}\right)\tau, \ x = L\xi$$

we have

Lesson 22: Dimensionless Problems

$$u_t = u_\tau \tau_t = \left(\frac{\alpha}{L}\right) u_\tau$$

$$u_{tt} = \frac{\partial}{\partial t}\left[\left(\frac{\alpha}{L}\right) u_\tau\right] = \left(\frac{\alpha}{L}\right) \frac{\partial}{\partial t}(u_\tau) = \left(\frac{\alpha}{L}\right)^2 u_{\tau\tau}$$

$$u_x = u_\xi \xi_x = \left(\frac{1}{L}\right) u_\xi$$

$$u_{xx} = \left(\frac{1}{L}\right)\frac{\partial}{\partial x}(u_\xi) = \left(\frac{1}{L}\right)\frac{\partial}{\partial \xi}(u_\xi)\xi_x =$$

Plugging the above derivatives in the PDE $u_{tt} = \alpha^2 u_{xx}$ yields the dimensionless form

$$\left(\frac{\alpha}{L}\right)^2 u_{\tau\tau} = \alpha^2 \left(\frac{1}{L}\right)^2 u_{\xi\xi} \Rightarrow u_{\tau\tau} = u_{\xi\xi}$$

The BCs and ICs transform in the same way by simply substituting for x and t. Doing this we get the PDE with the same dependent variable u but new dimensionless independent variables ξ, τ:

BC: $\begin{cases} u(0,\tau) = 0 \\ u(1,\tau) = 0 \end{cases} \quad 0 < \tau < \infty$

IC: $\begin{cases} u(\xi, 0) = \sin(\pi\xi) + \dfrac{1}{2}\sin(3\pi\xi) \\ u_\tau(\xi, 0) = 0 \end{cases} \quad 0 \le \xi \le 1$

Note: In the above BCs and ICs we simply replace x, t by ξ, τ although ξ ranges from 0 to 1, and although τ ranges from 0 to ∞ the same as t, they do not change at the same rate.

144   Lesson 22: Dimensionless Problems

> 2. Find the dimensionless formation of the problem
>
> $$\text{PDE: } u_t = \alpha^2 u_{xx} \qquad 0 < x < L,\ 0 < t < \infty$$
>
> $$\text{BC: } \begin{cases} u(0,t) = T_1 \\ u(L,t) = 0 \end{cases} \qquad 0 < t < \infty$$
>
> $$\text{IC: } \begin{cases} u(x,0) = \sin\left(\dfrac{\pi x}{L}\right) + \dfrac{1}{2}\sin\left(\dfrac{3\pi x}{L}\right) \\ u_t(x,0) = T_2 \end{cases} \qquad 0 \le x \le L$$

Solution: In this problem we want to transform both the dependent variable u and the independent variables x, t to dimensionless form. We do this by transforming the dependent variable by

$$U(x,t) = \frac{1}{T_1} u(x,t)$$

and the independent variables by

$$\tau = \left(\frac{\alpha}{L}\right)^2 t, \quad \xi = \left(\frac{1}{L}\right) x$$

Note that the coefficient of t, which is $(\alpha/L)^2$, has units of 1/time, hence the new time τ is dimensionless. Taking the required derivatives with the help of the functional diagram in Figure 22.1

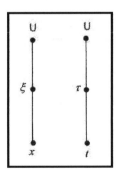

Figure 22.1 Functional diagram for computing partial derivatives

Lesson 22: Dimensionless Problems

we find

$$u_t = T_1 U_t = T_1\left[\frac{\partial U}{\partial \tau}\frac{d\tau}{dt}\right] = T_1\left[\left(\frac{\alpha}{L}\right)^2 U_\tau\right]$$

$$u_x = T_1 U_x = T_1\left[\frac{\partial U}{\partial \xi}\frac{d\xi}{dx}\right] = T_1\left[\left(\frac{1}{L}\right)U_\xi\right]$$

$$u_{xx} = T_1 U_{xx} = T_1\left[\frac{\partial}{\partial x}(U_x)\right] = T_1\left[\frac{\partial}{\partial x}\left(\frac{1}{L}\right)U_\xi\right] = T_1\left[\frac{1}{L}U_{\xi\xi}\frac{d\xi}{dx}\right] = T_1\left[\left(\frac{1}{L}\right)^2 U_{\xi\xi}\right]$$

Plugging these derivatives in the PDE $u_t = \alpha^2 u_{xx}$, we find

$$T_1\left[\left(\frac{\alpha}{L}\right)^2 U_\tau\right] = \alpha^2 T_1\left[\left(\frac{1}{L}\right)^2 U_{\xi\xi}\right]$$

and after cancelation, we get the pure dimensionless PDE

$$U_\tau = U_{\xi\xi} \quad 0 < \xi < 1, \ 0 < \tau < \infty$$

Hence, we arrive at the new dimensionless IBVP

PDE: $U_\tau = U_{\xi\xi}$ $\quad 0 < \xi < 1, \ 0 < \tau < \infty$

BC: $\begin{cases} U(0,\tau) = 1 \\ U(1,\tau) = 0 \end{cases}$ $\quad 0 < \tau < \infty$

IC: $U(\xi, 0) = T_2 / T_2$ $\quad 0 \le \xi \le 1$

Note that the new space variable ξ doesn't have units of inches, feet, cm, meters, ... , it has *no* units, it's just a variable. The same holds for the time variable τ, it doesn't have units of seconds, hours, days, ... it is just a time variable with *no* units.

Lesson 22: Dimensionless Problems

3. Transform the IBVP

PDE: $u_t = \alpha^2 u_{xx}$ $0 < x < L,\ 0 < t < \infty$

BC: $\begin{cases} u(0,t) = T_1 \\ u(L,t) = T_2 \end{cases}$ $0 < t < \infty$

IC: $u(x,0) = \sin\left(\dfrac{\pi x}{L}\right)$ $0 \le x \le L$

to

PDE: $U_t = \alpha^2 U_{xx}$ $0 < x < L,\ 0 < t < \infty$

BC: $\begin{cases} U(0,t) = 0 \\ U(L,t) = 1 \end{cases}$ $0 < t < \infty$

IC: $U(x,0) = \dfrac{\sin(\pi x/L) - T_1}{T_2 - T_1}$ $0 \le x \le L$

Solution: Making the substitution

$$u(x,t) = T_1 + (T_2 - T_1)U(x,t)$$

we have

$$u_t = U_t$$
$$u_x = U_{xx}$$
$$u(0,t) = T_1 + (T_2 - T)U(0,t)$$
$$u(1,t) = T_1 + (T_2 - T)U(1,t)$$
$$u(x,0) = T_1 + (T_2 - T)U(x,0)$$

and plugging these values in the IBVP

Lesson 22: Dimensionless Problems

PDE: $u_t = \alpha^2 u_{xx}$ $\quad 0 < x < L,\ 0 < t < \infty$

BC: $\begin{cases} u(0,t) = T_1 \\ u(L,t) = T_2 \end{cases}$ $\quad 0 < t < \infty$

IC: $u(x,0) = \sin\left(\dfrac{\pi x}{L}\right)$ $\quad 0 \le x \le L$

yields

PDE: $U_t = \alpha^2 U_{xx}$ $\quad 0 < x < L,\ 0 < t < \infty$

BC: $\begin{cases} T_1 + (T_2 - T_1)U(0,t) = T_1 \\ T_1 + (T_2 - T_1)U(1,t) = T_2 \end{cases}$ $\quad 0 < t < \infty$

IC: $T_1 + (T_2 - T_1)U(x,0) = \sin\left(\dfrac{\pi x}{L}\right)$ $\quad 0 \le x \le L$

or simplifying, we have the desired IBVP

PDE: $U_t = \alpha^2 U_{xx}$ $\quad 0 < x < L,\ 0 < t < \infty$

BC: $\begin{cases} U(0,t) = 0 \\ U(L,t) = 1 \end{cases}$ $\quad 0 < t < \infty$

IC: $U(x,0) = \dfrac{\sin(\pi x/L) - T_1}{T_2 - T_1}$ $\quad 0 \le x \le L$

---

4. Can you think of a physical reason why the new time variable $\tau = ct$ would eliminate the parameter c^2 in the wave equation

$$u_{tt} = c^2 u_{xx}$$

Remember what α means in terms of the velocity of the wave. Intuition plays a big role in finding the most desirable new coordinates.

Solution: With respect to time scale t velocity of the wave is c and so with respect to the time scale $\tau = ct$ the velocity of the wave is 1.

> 5. How would you pick a new space variable ξ so that the equation
>
> $$u_t + v u_x = 0$$
>
> is transformed to dimensionless form
>
> $$u_t + u_\xi = 0$$

Solution: Trying the new variable $\xi = kx$, we have

$$u_x = u_\xi \xi_x = k u_\xi$$

and so if we pick $k = 1/v$ (i.e. $\xi = x/v$), we get the equation

$$u_t + v u_x = u_t + v\left(\frac{1}{v}\right) u_\xi = u_t + u_\xi = 0$$

ΟΠΘΩΞ

Lesson 23: Classification of PDEs (Canonical Form of the Hyperbolic Equation)

1. State whether the following PDEs are hyperbolic, parabolic, or elliptic?

   a) $u_{xx} - u_{xy} = 0$

   b) $u_{tt} = u_{xx} + u_x + hu$

   c) $u_{xx} + 3u_{yy} = \sin x$

   d) $u_{xx} + u_{yy} = f(x,t)$

   e) $u_{rr} + \frac{1}{r}u_r + \frac{1}{r^2}u_{\theta\theta} = f(r,\theta)$

Solution: The general second-order, linear PDE in two independent variables is

$$Au_{xx} + Bu_{xy} + Cu_{yy} + Du_x + Eu_y + Fu = G$$

where A, B, C, D, E, F, G are functions of x, y (where of course constants are also functions of x, y). The equation is classified as elliptic, parabolic, or hyperbolic according to

$$B^2 - 4AC < 0 \Rightarrow \text{equation is elliptic}$$
$$B^2 - 4AC = 0 \Rightarrow \text{equation is parabolic}$$
$$B^2 - 4AC > 0 \Rightarrow \text{equation is hyperbolic}$$

   a) $A=1, B=-1, C=0 \Rightarrow B^2 - 4AC = 1 \Rightarrow$ hyperbolic
   b) $A=1, B=0, C=-1 \Rightarrow B^2 - 4AC = 4 \Rightarrow$ hyperbolic
   c) $A=1, B=0, C=3 \Rightarrow B^2 - 4AC = -12 \Rightarrow$ elliptic
   d) $A=1, B=0, C=1 \Rightarrow B^2 - 4AC = -4 \Rightarrow$ elliptic
   e) $A=1, B=0, C=1/r^2 \Rightarrow B^2 - 4AC = -4/r^2 \Rightarrow$ elliptic

2. Verify equations $(23.4), (23.5)$ and (23.6) in the text.

Solution: Use the diagram in Figure 23.2

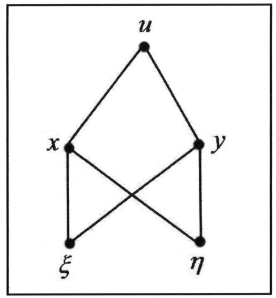

Figure 23.2 Functional diagram

To compute the required (messy) partial derivatives like you learned how to do in multivariable calculus.

---

3. Verify that the equation

$$3u_{xx} + 7u_{xy} + 2u_{yy} = 0$$

is hyperbolic for all x, y and find the new characteristic coordinates.

Solution: We have

$$A = 3, B = 7, C = 2 \quad \Rightarrow \quad B^2 - 4AC = 25 > 0$$

and so the equation is hyperbolic for all x, y. The characteristic equations are

$$\frac{dy}{dx} = \frac{B - \sqrt{B^2 - 4AC}}{2A} = \frac{7 - \sqrt{49 - 24}}{6} = \frac{1}{3}$$

$$\frac{dy}{dx} = \frac{B + \sqrt{B^2 - 4AC}}{2A} = \frac{7 + \sqrt{49 - 24}}{6} = 2$$

Lesson 23: Classification of PDEs (Hyperbolic Equation)

Hence

$$y = \frac{1}{3}x + c_1$$
$$y = 2x + c_2$$

and so the characteristic coordinates ξ, η for the canonical form are

$$\xi = y - \frac{1}{3}x$$
$$\eta = y - 2x$$

---

4. Continue Problem 3 by finding the canonical form of

$$3u_{xx} + 7u_{xy} + 2u_{yy} = 0$$

which has the form

$$u_{\xi\eta} = \Psi(\xi, \eta, u, u_\xi, u_\eta)$$

Solution: We see that $A = 3, B = 7, C = 2, D = E = F = G = 0$ and so the characteristic equations are

$$\frac{dy}{dx} = \frac{B - \sqrt{B^2 - 4AC}}{2A} = \frac{7 - \sqrt{49 - 24}}{6} = \frac{1}{3}$$

$$\frac{dy}{dx} = \frac{B + \sqrt{B^2 - 4AC}}{2A} = \frac{7 + \sqrt{49 - 24}}{6} = 2$$

Hence

$$y = \frac{1}{3}x + c_1$$
$$y = 2x + c_2$$

We now find the new canonical coordinates ξ, η by solving the above equations for c_1, c_2, or

$$\xi = y - \frac{1}{3}x$$
$$\eta = y - 2x$$

We now find the canonical equation
$$\bar{A}u_{\xi\xi} + \bar{B}u_{\xi\eta} + \bar{C}u_{\eta\eta} + \bar{D}u_\xi + \bar{E}u_\eta + Fu = \bar{G}$$
where

$$\bar{A} = 3\xi_x^2 + 7\xi_x\xi_y + 2\xi_y^2$$
$$\bar{B} = 6\xi_x\eta_x + 7(\xi_x\eta_y + \xi_y\eta_x) + 4\xi_y\eta_y$$
$$\bar{C} = 3\eta_x^2 + 7\eta_x\eta_y + 2\eta_y^2$$
$$\bar{D} = 3\xi_{xx} + 7\xi_{xy} + 2\xi_{yy}$$
$$\bar{E} = 3\eta_{xx} + 7\eta_{xy} + 2\eta_{yy}$$
$$\bar{F} = F = 0$$
$$\bar{G} = G = 0$$

and computing the derivatives
$$\xi_x = -1/3,\ \xi_y = 1,\ \eta_x = -2,\ \eta_y = 1$$
we find
$$\bar{A} = 0,\ \bar{B} = -25/3,\ \bar{C} = 0,\ \bar{D} = \bar{E} = \bar{F} = \bar{G} = G = 0$$

Hence, the canonical equation is
$$-\frac{25}{3}u_{\xi\eta} = 0$$
or simply
$$u_{\xi\eta} = 0$$

Lesson 23: Classification of PDEs (Hyperbolic Equation)

5. Continue Problem 4 by finding the alternative canonical form

$$u_{\alpha\alpha} - u_{\beta\beta} = \Psi(\alpha,\beta,u,u_\alpha,u_\beta)$$

Solution Given a hyperbolic PDE in canonical form $u_{\xi\eta} = \Phi(\xi,\eta,u,u_\xi,u_\eta)$, it can be written in an alternate canonical form

$$u_{\alpha\alpha} - u_{\beta\beta} = \Psi(\alpha,\beta,u,u_\alpha,u_\beta)$$

using the transformation
$$\alpha = \xi + \eta$$
$$\beta = \xi - \eta$$

and with the help of the diagram in Figure 23.1 we compute the following partial derivatives

$$u_\xi = u_\alpha \alpha_\xi + u_\beta \beta_\xi = u_\alpha + u_\beta$$
$$u_\eta = u_\alpha \alpha_\eta + u_\beta \beta_\eta = u_\alpha - u_\beta$$

$$u_{\xi\eta} = \frac{\partial}{\partial \xi}(u_\eta) = \frac{\partial}{\partial \xi}(u_\alpha - u_\beta) = u_{\alpha\alpha}\alpha_\xi + u_{\alpha\beta}\beta_\xi - u_{\alpha\beta}\alpha_\xi - u_{\beta\beta}\beta_\xi$$
$$= u_{\alpha\alpha} + u_{\alpha\beta} - u_{\alpha\beta} - u_{\beta\beta}$$
$$= u_{\alpha\alpha} - u_{\beta\beta}$$

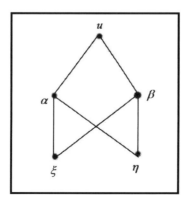

Figure 23.1 Diagram illustrating functional dependence

In Problem 4 we had $u_{\xi\eta} = 0$ and since $u_{\alpha\alpha} - u_{\beta\beta} = u_{\xi\eta}$ the alternate canonical for is

$$u_{\alpha\alpha} - u_{\beta\beta} = 0$$

---

6. Find the new characteristic coordinates for

$$u_{xx} + 4u_{yy} = 0$$

Solve the transformed equation in the new coordinates and then transform back to solve the original problem in the original coordinates.

Solution: We see that $A = 1, B = 0, C = 4, D = E = F = G = 0$ and so the characteristic equations are

$$\frac{dy}{dx} = \frac{B - \sqrt{B^2 - 4AC}}{2A} = \frac{0 - \sqrt{0 - 16}}{2} = -2i$$

$$\frac{dy}{dx} = \frac{B + \sqrt{B^2 - 4AC}}{2A} = \frac{0 + \sqrt{0 - 16}}{2} = 2i$$

Hence

$$y = -2ix + c_1$$
$$y = 2ix + c_2$$

and so the new coordinates ξ, η for the canonical form are

$$\xi = y + 2ix$$
$$\eta = y - 2ix$$

The original equation is elliptic, hence complex characteristic coordinates.

$$\Sigma\Pi\Psi Z\Omega\Sigma$$

Lesson 24: The Wave Equation in Two and Three Dimensions (Free Space)

> 1. Show that in one dimension, we can find the solution of
>
> PDE: $u_{tt} = c^2 u_{xx}$ $\quad -\infty < x < \infty,\ 0 < t < \infty$
>
> ICs: $\begin{cases} u(x,0) = \phi(x) \\ u_t(x,0) = 0 \end{cases}$ $\quad -\infty < x < \infty$
>
> by differentiating (with respect to t) the solution of
>
> PDE: $u_{tt} = c^2 u_{xx}$ $\quad -\infty < x < \infty,\ 0 < t < \infty$
>
> ICs: $\begin{cases} u(x,0) = 0 \\ u_t(x,0) = \phi(x) \end{cases}$ $\quad -\infty < x < \infty$

Solution: The solution of the IVP when the ICs are

ICs: $\begin{cases} u(x,0) = 0 \\ u_t(x,0) = \phi(x) \end{cases}$ $\quad -\infty < x < \infty$

is

$$u(x,t) = \frac{1}{2c} \int_{x-ct}^{x+ct} \phi(s)\,dx$$

If we differentiate this function with respect to t using Leibniz rule

$$\frac{d}{dt} \int_{f(t)}^{g(t)} \phi(s)\,ds = \phi[g(t)]g'(t) - \phi[f(t)]f'(t)$$

we get

$$\frac{du}{dt} = \frac{1}{2c}\left[\phi(x+ct)\frac{d}{dt}(x+ct) - \phi(x-ct)\frac{d}{dt}(x-ct)\right]$$

$$= \frac{1}{2}[\phi(x+ct) + \phi(x-ct)]$$

which is the solution when the IC are

Lesson 24: The Wave Equation in Two and Three Dimensions

ICs: $\begin{cases} u(x,0) = \phi(x) \\ u_t(x,0) = 0 \end{cases}$ $-\infty < x < \infty$

2. Apply the result from Problem 1 to find the solution of

PDE: $u_{tt} = u_{xx}$ $-\infty < x < \infty, \ 0 < t < \infty$

ICs: $\begin{cases} u(x,0) = x \\ u_t(x,0) = 0 \end{cases}$ $-\infty < x < \infty$

Solution: We start with the solution of the IVP

PDE: $u_{tt} = u_{xx}$ $-\infty < x < \infty, \ 0 < t < \infty$

ICs: $\begin{cases} u(x,0) = 0 \\ u_t(x,0) = x \end{cases}$ $-\infty < x < \infty$

which is

$$u(x,t) = \frac{1}{2} \int_{x-t}^{x+t} \phi(s) \, ds$$

$$= \frac{1}{2} \int_{x-t}^{x+t} s \, ds$$

$$= \frac{1}{2} \left[\frac{1}{2} s^2 \Big|_{x-t}^{x+t} \right]$$

$$= \frac{1}{4} \left[(x+t)^2 - (x-t)^2 \right]$$

$$= xt$$

Now if we differentiate this solution with respect to t we get x. Now if we find the solution of the IVP with IC

ICs: $\begin{cases} u(x,0) = x \\ u_t(x,0) = 0 \end{cases}$ $-\infty < x < \infty$

Lesson 24: The Wave Equation in Two and Three Dimensions

we get the same result

$$u(x,t) = \frac{1}{2}\left[\phi(x+t) + \phi(x-t)\right]$$
$$= \frac{1}{2}\left[(x+t) + (x-t)\right]$$
$$= x$$

---

3. Illustrate by picture and words the spherical wave solution of the three-dimensional problem

PDE: $u_{tt} = c^2 \nabla^2 u \quad (x,y,z) \in \mathfrak{R}^3$

ICs: $\begin{cases} u(x,y,z,0) = 0 \\ u_t(x,y,z,0) = \begin{cases} 1 & x^2 + y^2 + z^2 \leq 1 \\ 0 & \text{elsewhere} \end{cases} \end{cases}$

Solution: Since the solution of the above problem is

$$u(x,y,z,t) = t\bar{\psi}$$

one can see that the initial disturbance region (which is the unit sphere) gives rise to an outwardly propagating annulus. Each point (x,y,z) outside the unit sphere will suddenly experience a disturbance and will be disturbed for a length of time after which $u(x,y,z,t) = 0$.

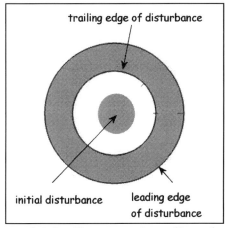

Figure 24.1 Propagating disturbance

Lesson 24: The Wave Equation in Two and Three Dimensions

4. (Worked in conjunction with Problem 3) What is the two-dimensional solution of the analogous cylindrical problem

PDE: $u_{tt} = c^2 \nabla^2 u \quad (x,y) \in \Re^2$

ICs: $\begin{cases} u(x,y,0) = 0 \\ u_t(x,y,0) = \begin{cases} 1 & x^2 + y^2 \leq 1 \\ 0 & \text{elsewhere} \end{cases} \end{cases}$

Solution: For this problem each point (x,y) outside the unit circle will suddenly experience a shock wave and as time gets large the solution $u(x,y,t)$ gradually goes to zero. The solution would look more or less like a circular wave after dropping a pebble in a lake.

5. (Worked in conjunction with Problem 3) What is the one-dimensional solution of the plane wave equation

PDE: $u_{tt} = c^2 u_{xx} \quad -\infty < x < \infty$

ICs: $\begin{cases} u(x,0) = 0 \\ u_t(x,0) = \begin{cases} 1 & |x| \leq 1 \\ 0 & \text{elsewhere} \end{cases} \end{cases}$

Solution: We have seen in the lesson that the solution will be

$$u(x,t) = \begin{cases} 0 & -\infty < x < -ct-1 \\ \dfrac{1+x+ct}{2c} & -ct-1 \leq x < -ct+1 \\ \dfrac{1}{c} & -ct+1 \leq x < ct-1 \\ \dfrac{1-x+ct}{2c} & ct-1 \leq x < ct+1 \\ 0 & ct+1 \leq x < \infty \end{cases}$$

This solution is illustrated in Figure 18.4 in the text.

Lesson 24: The Wave Equation in Two and Three Dimensions 159

> **6.** What is the physical interpretation of why Huygen's principle does not hold in two dimensions? with respect to t.

Solution: Imagine a single initial disturbance at the origin $(0,0)$. Since Huygen's principle does not holds in 2 dimensions this means that after a given length of time any other point (x,y) will suddenly experience a shock due to this disturbance and then gradually die out as time gets large. The reason the shock gradually goes to zero is because the origin in two dimensions can physically be interpreted as an infinite line perpendicular to the sy plane and the points in the xy plane are in reality being disturbed by all the points along this line. As time increases all points along this line will be disturbing the point (x,y) and as time increases it is the points further away that are doing the disturbing.

> **7.** Use Leibniz's rule
>
> $$\frac{d}{dt}\int_{f(t)}^{g(t)} F(\xi,t)\,d\xi = \int_{f(t)}^{g(t)} \frac{\partial F}{\partial t}(\xi,t)\,dt + g'(x)F[g(t),t] - F[f(t),t]f'(t)$$
>
> to differentiate the followng integral with respect to t.
>
> $$u(x,t) = \frac{1}{2c}\int_{x-ct}^{x+ct} \phi(\xi)\,d\xi$$

Solution: Leibniz rule

$$\frac{d}{dx}\int_{f(t)}^{g(t)} F(\xi,t)\,d\xi = \int_{f(t)}^{g(t)} \frac{\partial F}{\partial t}(\xi,t)\,dt + g'(x)F[g(t),t] - F[f(t),t]$$

tells us how to differentiate an integral with respect to t where the limits and the integrand can depend on t. Direct application of the formula gives

$$\frac{d}{dt}\int_{x-ct}^{x+ct} \phi(s)\,ds = \phi(c+t)\frac{\partial}{\partial t}(x+ct) - \phi(x-ct)\frac{\partial}{\partial t}(x-ct)$$
$$= c[\phi(x+ct) + \phi(x-ct)]$$

ΟΠΘΡΣΨ

Lesson 25: The Finite Fourier Transforms (Sine and Cosine Transforms)

1. Solve the diffusion following diffusion problem with insulated boundaries, given by

   PDE: $u_t = u_{xx}$ $0 < x < 1,\ 0 < t < \infty$

   BCs: $\begin{cases} u_x(0,t) = 0 \\ u_x(1,t) = 0 \end{cases}$ $0 < t < \infty$

   ICs: $u(x,0) = 1 + \cos(\pi x) + \dfrac{1}{2}\cos(3\pi x)$ $0 \leq x \leq 1$

Solution: Applying the cosine transform $C_n = C[u]$, we get

$$\frac{dC_n}{dt} + (n\pi)^2 C_n(t) = 0$$

$$C_n(0) = \begin{cases} 1 & n = 0, 1 \\ 0 & n = 2, 4, 5, \ldots \\ 1/2 & n = 3 \end{cases}$$

Solving these equations, we get

$$C_n(t) = \begin{cases} 1 & n = 0 \\ e^{-\pi^2 t} & n = 1 \\ 0 & n = 2 \\ \dfrac{1}{2} e^{-(3\pi)^2 t} & n = 3 \end{cases}$$

Hence, we have

$$u(x,t) = 1 + e^{-\pi^2 t}\cos(\pi x) + \frac{1}{2} e^{-(3\pi)^2 t}\cos(3\pi x)$$

Lesson 25: The Finite Fourier Transform

2. Solve the general IBVP

PDE: $u_t = \alpha^2 u_{xx} + bu + f(x,t)$ $0 < x < 1, \ 0 < t < \infty$

BCs: $\begin{cases} u(0,t) = 0 \\ u(1,t) = 0 \end{cases}$ $0 < t < \infty$

ICs: $u(x,0) = 0$ $0 \le x \le 1$

Solution: Taking the sine transform $S_n = S[u]$ of the IBVP, we get

ODE: $\dfrac{dS_n}{dt} + \left[(n\pi\alpha)^2 + b\right] S_n = F_n(t)$

IC: $S_n(0) = 0$

Solving this ODE by finding an integrating factor, we get

$$S_n(t) = e^{-\left[(n\pi\alpha)^2 + b\right]t} \int_0^t e^{\left[(n\pi\alpha)^2 + b\right]s} F_n(s)\,ds$$

Hence, the solution is

$$u(x,t) = \sum_{n=1}^{\infty} S_n(t) \sin(n\pi x)$$

3. Derive the basic laws for the Sine and Cosine transforms. Can you see why it would be hard to solve a differential equation that contained the term u_t?

$$S_n[u_t] = \frac{dS_n[u]}{dt}$$

$$S_n[u_{tt}] = \frac{d^2 S_n[u]}{dt^2}$$

$$S[u_{xx}] = -[n\pi L]^2 S[u] + \frac{2n\pi}{L^2}\left[u(0,t) + (-1)^{n+1} u(L,0)\right]$$

$$C[u_{xx}] = -[n\pi L]^2 C[u] - \frac{2}{L^2}\left[u_x(0,t) + (-1)^{n+1} u_x(L,0)\right]$$

Solution: The Sine and Cosine transforms of a function $u(x,t)$ are defined by

$$S_n[u] = \frac{2}{L}\int_0^L u(x,t)\sin\left(\frac{n\pi x}{L}\right)dx$$

$$C_n[u] = \frac{2}{L}\int_0^L u(x,t)\cos\left(\frac{n\pi x}{L}\right)dx$$

a) The law

$$S_n[u_t] = \frac{dS_n[u]}{dt}$$

which says

$$\frac{2}{L}\int_0^L u_t(x,t)\sin\left(\frac{n\pi x}{L}\right)dx = \frac{d}{dt}\left\{\frac{2}{L}\int_0^L u(x,t)\sin\left(\frac{n\pi x}{L}\right)dx\right\}$$

is a restatement of the following general law of calculus

$$\frac{d}{dt}\int_a^b f(t,s)\,ds = \int_a^b \frac{\partial f}{\partial t}(t,s)\,ds$$

which allows one to differentiate (as long as the function is differentiable of course) inside the integral. The law is called Leibniz rule in calculus.

b) To prove

$$S_n[u_{tt}] = \frac{d^2 S_n[u]}{dt^2}$$

we must show

$$\frac{2}{L}\int_0^L u_{tt}(x,t)\sin\left(\frac{n\pi x}{L}\right)dx = \frac{d^2}{d^2 t}\left\{\frac{2}{L}\int_0^L u(x,t)\sin\left(\frac{n\pi x}{L}\right)dx\right\}$$

Again, this statement is a special case of Leibniz rule that lets one differentiate a function inside an integral.

c) To prove

$$S[u_{xx}] = -[n\pi L]^2 S[u] + \frac{2n\pi}{L^2}\left[u(0,t) + (-1)^{n+1} u(L,0)\right]$$

Lesson 25: The Finite Fourier Transform

we must show

$$\frac{2}{L}\int_0^L u_{xx}(x,t)\sin\left(\frac{n\pi x}{L}\right)dx = -[n\pi L]^2\left\{\frac{2}{L}\int_0^L u(x,t)\sin\left(\frac{n\pi x}{L}\right)ds\right\} + \frac{2n\pi}{L^2}\left[u(0,t)+(-1)^{n+1}u(L,0)\right]$$

If you look carefully you will see that this law is glorified version of the integration by parts law

$$\int_a^b U(x)V'(x)dx = \left[U(b)V(b)-U(a)V(a)\right] - \int_a^b V(x)U'(x)dx$$

except in our case we have functions of two variables not one. So letting

$$U = \sin\left(\frac{n\pi x}{L}\right), \quad dV = u_{xx}$$

and using the integration by parts formula, we get

$$\frac{2}{L}\int_0^L u_{xx}(x,t)\sin\left(\frac{n\pi x}{L}\right)dx = \frac{2}{L}u_x(x,t)\sin\left(\frac{n\pi x}{L}\right)\Big|_0^L - \frac{2}{L}\left(\frac{n\pi}{L}\right)\int_0^L u_x(x,t)\cos\left(\frac{n\pi x}{L}\right)dx$$

$$= -\left(\frac{2n\pi}{L^2}\right)\int_0^L u_x(x,t)\cos\left(\frac{n\pi x}{L}\right)dx$$

Doing this again, we let

$$U = \cos\left(\frac{n\pi x}{L}\right), \quad dV = u_x$$

we get

$$-\left(\frac{2n\pi}{L^2}\right)\int_0^L u_x(x,t)\cos\left(\frac{n\pi x}{L}\right)dx = -\left(\frac{2n\pi}{L^2}\right)\left[u(x,t)\cos\left(\frac{n\pi x}{L}\right)\right]_0^L - \left(\frac{2n\pi}{L^2}\right)\left(\frac{n\pi}{L}\right)\int_0^L u(x,t)\sin\left(\frac{n\pi x}{L}\right)dx$$

$$= -[n\pi L]^2 S[u] + \frac{2n\pi}{L^2}\left[u(0,t)+(-1)^{n+1}u(L,0)\right]$$

b) The law

$$C[u_{xx}] = -[n\pi L]^2 C[u] - \frac{2}{L^2}\left[u_x(0,t)+(-1)^{n+1}u_x(L,0)\right]$$

is another fun integration by parts we leave for the reader.

4. What is the finite sine transform of

$$f(x) = \sin(\pi x) + \frac{1}{2}\sin(3\pi x)$$

Graph the sine transform with $L=1$.

Solution: When the function is a sum of sine functions, we can simply pick of the coefficients, that is

$$S_n = \begin{cases} 1 & n=1 \\ 1/2 & n=3 \\ 0 & \text{all other } n's \end{cases}$$

5. What is the finite cosine transform of

$$f(x) = x, \ 0 \le x \le 1$$

What would the graph of the inverse transform look like for all values of x? You know it reproduces $f(x) = x, \ 0 \le x \le 1$ but what would it look like outside the interval $[0,1]$?

Solution: The finite cosine transform of the function $f(x) = x, \ 0 \le x \le 1$ of a *single* variable gives the coefficients in the Fourier cosine series of the function. Finding it we get

$$C_n[f] = \frac{2}{L}\int_0^L f(x)\cos(n\pi x)\,dx = 2\int_0^1 x\cos(n\pi x)\,dx = \begin{cases} \dfrac{1}{2} & n=0 \\ \dfrac{-4}{n^2\pi^2} & n=1,3, \\ 0 & n=2,4,\ldots \end{cases}$$

Hence the cosine series for $f(x) = x, \ 0 \le x \le 1$ is

Lesson 25: The Finite Fourier Transform

$$x = \frac{1}{2} - \frac{4}{\pi^2}\left[\cos(\pi x) + \frac{1}{9}\cos(3\pi x) + \frac{1}{25}\cos(5\pi x) + \cdots\right], \quad 0 \le x \le 1$$

Figures 25.1 a) and 25.1 b) shows graphs of the function $f(x) = x$, $0 \le x \le 1$ and the first 2 and 3 terms of the Fourier Cosine series. You can see how accurate the series approximates the function with only a couple terms.

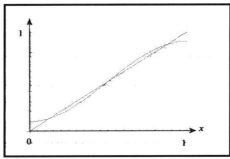

Figure 25.1 a) Graphs of $f(x) = x$ and $\dfrac{1}{2} - \dfrac{4}{\pi^2}\cos(\pi x)$

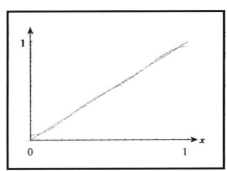

Figure 25.1 b) Graphs of $f(x) = x$ and $\dfrac{1}{2} - \dfrac{4}{\pi^2}\cos(\pi x) + \dfrac{1}{9}\cos(3\pi x)$

---

6. Solve the IBVP

    PDE: $u_t = u_{xx} + \sin(3\pi x) \quad 0 < x < 1, \; 0 < t < \infty$

    BCs: $\begin{cases} u(0,t) = 0 \\ u(1,t) = 0 \end{cases} \quad 0 < t < \infty$

    ICs: $u(x,0) = \sin(\pi x) \quad 0 \le x \le 1$

Solution: Taking the sine transform $S_n = S[u]$, we get

$$\text{ODE:} \quad \frac{dS_n}{dt} + (n\pi)^2 S_n = \begin{cases} 1 & n=3 \\ 0 & n \neq 3 \end{cases}$$

$$\text{IC:} \quad S_n(0) = \begin{cases} 1 & n=1 \\ 0 & n=2,3,4,\ldots \end{cases}$$

Writing these equations separately, we have

$n=1$
$$\text{ODE:} \quad \frac{dS_1}{dt} + \pi^2 S_1 = 0 \quad \Rightarrow S_1(t) = e^{-\pi^2 t}$$
$$\text{IC:} \quad S_1(0) = 1$$

$n=3$
$$\text{ODE:} \quad \frac{dS_3}{dt} + (3\pi)^2 S_3 = 1 \quad \Rightarrow S_3(t) = \frac{1}{(3\pi)^2}\left(1 - e^{-(3\pi)^2 t}\right)$$
$$\text{IC:} \quad S_3(0) = 0$$

$n \neq 1,3$
$$\text{ODE:} \quad \frac{dS_n}{dt} + (n\pi)^2 S_n = 0 \quad \Rightarrow S_n(t) = 0$$
$$\text{IC:} \quad S_n(0) = 0$$

Hence, we have

$$u(x,t) = \sum_{n=1}^{\infty} S_n(t) \sin(n\pi x)$$

$$= e^{-\pi^2 t} \sin(\pi x) + \frac{1}{(3\pi)^2}\left(1 - e^{-(3\pi)^2 t}\right)\sin(3\pi x)$$

You can check to see that this function satisfies the IBVP.

$$\text{NO}\Sigma\Psi\Xi\text{Z}$$

Lesson 26: Superposition (The Backbone of Linear Systems)

1. Show that if u_1 and u_2 are solutions to the following problems P_1 and P_2

P_1
PDE: $u_t = u_{xx} + \sin(\pi x)$ $0 < x < 1,\ 0 < t < \infty$

BCs: $\begin{cases} u(0,t) = 0 \\ u(1,t) = 0 \end{cases}$ $0 < t < \infty$

ICs: $u(x,0) = 0$ $0 \le x \le 1$

P_2
PDE: $u_t = u_{xx}$ $0 < x < 1,\ 0 < t < \infty$

BCs: $\begin{cases} u(0,t) = 0 \\ u(1,t) = 0 \end{cases}$ $0 < t < \infty$

ICs: $u(x,0) = \sin(2\pi x)$ $0 \le x \le 1$

respectively, then $u_1 + u_2$ satisfies the following problem P

P
PDE: $u_t = u_{xx} + \sin(\pi x)$ $0 < x < 1,\ 0 < t < \infty$

BCs: $\begin{cases} u(0,t) = 0 \\ u(1,t) = 0 \end{cases}$ $0 < t < \infty$

ICs: $u(x,0) = \sin(2\pi x)$ $0 \le x \le 1$

Solution: Since u_1 satisfies P_1 and u_2 satisfies P_2, we have

P_1
PDE: $\dfrac{\partial u_1}{\partial t} = \dfrac{\partial^2 u_1}{\partial x^2} + \sin(\pi x)$ $0 < x < 1,\ 0 < t < \infty$

BCs: $\begin{cases} u_1(0,t) = 0 \\ u_1(1,t) = 0 \end{cases}$ $0 < t < \infty$

ICs: $u_1(x,0) = 0$ $0 \le x \le 1$

Lesson 26: Superposition (The Backbone of Linear Systems)

$$P_2 \quad \begin{aligned} \text{PDE:} & \quad \frac{\partial u_2}{\partial t} = \frac{\partial^2 u_2}{\partial x^2} \quad 0<x<1, \ 0<t<\infty \\ \text{BCs:} & \quad \begin{cases} u_2(0,t)=0 \\ u_2(1,t)=0 \end{cases} \quad 0<t<\infty \\ \text{ICs:} & \quad u_2(x,0)=\sin(2\pi x) \quad 0\le x \le 1 \end{aligned}$$

If we now add the respective equations in the problems P_1 and P_2, we get that $u_1 + u_2$ satisfies problem P:

PDE: $\quad \dfrac{\partial(u_1+u_2)}{\partial t} = \dfrac{\partial^2(u_1+u_2)}{\partial x^2} + \sin(\pi x) \quad 0<x<1, \ 0<t<\infty$

BCs: $\quad \begin{cases} (u_1+u_2)(0,t)=0 \\ (u_1+u_2)(1,t)=0 \end{cases}$

ICs: $\quad (u_1+u_2)(x,0) = \sin(2\pi x) \quad 0 \le x \le 1$

---

2. Solve the following IBVP by superposition. Each sub problem can be solved in any manner you wish.

$$\begin{aligned} \text{PDE:} & \quad u_{tt} = u_{xx} + \sin(3\pi x) \quad 0<x<1, \ 0<t<\infty \\ \text{BCs:} & \quad \begin{cases} u(0,t)=0 \\ u(1,t)=0 \end{cases} \quad 0<t<\infty \\ \text{ICs:} & \quad \begin{cases} u(x,0)=\sin(\pi x) \\ u_t(x,0)=0 \end{cases} \quad 0 \le x \le 1 \end{aligned}$$

Solution The solution with homogeneous PDE and nonhomogeneous IC can be solved by separation of variables getting

$$u_1(x,t) = \cos(\pi t)\sin(\pi x)$$

The solution to the problem with nonhomogeneous PDE and homogeneous IC can be solved by the finite sine transform, getting

Lesson 26: Superposition (The Backbone of Linear Systems) 169

$$u_2(x,t) = \left[-\frac{1}{(3\pi)^2}\cos(3\pi t) - 1\right]\sin(3\pi x)$$

Hence, the solution of the original problem is the sum

$$u(x,t) = u_1(x,t) + u_2(x,t)$$

$$= \cos(\pi t)\sin(\pi x) - \left[\frac{1}{(3\pi)^2}\cos(3\pi t) + 1\right]\sin(3\pi x)$$

3. Suppose u_1 and u_2 are solutions to the following equations. For which equation is $u_1 + u_2$ a solution?

   a) $u_t = u_{xx}$
   b) $u_t = u_{xx} + e^t$
   c) $u_t = e^{-t}u_{xx}$
   d) $u_t = u_{xx} + u^2$

What conclusions can you reach from your answers?

Solution:

a) $u_1 + u_2$ is a solution since the equation is linear and homogeneous
b) $u_1 + u_2$ is not a solution since the equation is not homogeneous
c) $u_1 + u_2$ is a solution since the equation is linear and homogeneous
d) $u_1 + u_2$ is not a solution since the equation is not linear

4. Find four initial-value problems whose solutions sum to the solution of the following problem?

Lesson 26: Superposition (The Backbone of Linear Systems)

$$\text{PDE:} \quad u_t = u_{xx} + f(x,t) \quad 0 < x < 1, \ 0 < t < \infty$$

$$\text{BCs:} \quad \begin{cases} u(0,t) = g_1(t) \\ u(1,t) = g_2(t) \end{cases} \quad 0 < t < \infty$$

$$\text{ICs:} \quad u(x,0) = \phi(x) \quad 0 \le x \le 1$$

Solution: In each of the four sub problems let one of the right hand sides be non zero, the others zero. The sum of the four solutions will then be the solution of the given problem.

---

5. Solve the following IVP. Can you verify your answer?

$$\text{ODE:} \quad \frac{dU_n(t)}{dt} + (n\pi)^2 U_n = F_n(t)$$

$$\text{IC:} \quad U_n(0) = 0$$

Solution: This is a standard ODE with IC. One finds the general solution to the ODE by multiplying each side of the equation by the integrating factor method you learned in ODE class. Doing this we get the general solution of the IVP to be

$$U_n(t) = e^{-(n\pi)^2 t} \int_0^t e^{(n\pi)^2 s} F_n(s) \, ds$$

6. Suppose u_1 and u_2 both satisfy the following linear homogeneous BCs. Does the sum $u_1 + u_2$ satisfy the BCs?

Solution: Yes, this can be verified by direct substitution.

ΣΤΖΩΞΠ

Lesson 27: First-Order Equations (Method of Characteristics)

1. Solve the following simple convection problem. What does the solution look like?

PDE: $u_x + u_t = 0 \quad -\infty < x < \infty, \ 0 < t < \infty$

IC: $u(x,0) = \cos x \quad -\infty < x < \infty$

Solution: Using the method of characteristics we let

$$\frac{dx}{ds} = 1, \quad \frac{dt}{ds} = 1$$

which yield

$$x(s) = s + c_1$$
$$t(s) = s + c_2$$

Using IC

$$x(0) = \tau, \ t(0) = 0$$

we get $c_1 = \tau, c_2 = 0$ and so

$$x = s + \tau$$
$$t = s$$

If we eliminate s from the previous equations, we get the characteristic curves in the xt plane as a function of parameter τ, which are

$$x - t = \tau \ (-\infty < \tau < \infty)$$

which is a family of straight lines in the xt.

Now, we can write the PDE and IC in terms of s and τ as

ODE: $\dfrac{du}{ds} = 0 \quad 0 < s < \infty$

IC: $u(\tau, 0) = \cos \tau$

which has the solution

Lesson 27: First-Order Equations (Method of Characteristics)

$$u(s,\tau) = \sin\tau$$

or in terms of the original coordinates

$$u(x,t) = \cos(x-t)$$

Note that this solution is a moving cosine curve moving to the right with velocity one.

---

2. Solve the following convection problem. Note that the IC begins at $t=1$. What do the characteristics look like? Check your solution.

$$\text{PDE: } xu_x + tu_t + 2u = 0 \quad -\infty < x < \infty, \ 1 < t < \infty$$

$$\text{IC: } u(x,1) = \sin x \quad -\infty < x < \infty$$

Solution: Using the method of characteristics we let

$$\frac{dx}{ds} = x, \quad \frac{dt}{ds} = t$$

which yield

$$x = c_1 e^s$$

$$t = c_2 e^s$$

We now use the IC (note time starts at $t=1$ in this problem)

$$x(0) = \tau, \ t(0) = 1$$

we get $c_1 = \tau$, $c_2 = 1$ and so

$$x = \tau e^s$$

$$t = e^s$$

We no eliminate the variable s to get the characteristic curves, which are curves in the xt plane with parameter τ, and in this case they are

Lesson 27: First-Order Equations (Method of Characteristics)

$$\frac{x}{t} = \tau \quad (-\infty < \tau < \infty)$$

The PDE and IC can now be written as

ODE: $\quad \dfrac{du}{ds} + 2u = 0 \quad 0 < s < \infty$

IC: $\quad u(\tau, 0) = \sin \tau$

which has the solution

$$u(s, \tau) = (\sin \tau) e^{-2s}$$

or substituting in

$$\tau = \frac{x}{t}, \quad s = \ln t$$

we have

$$u(s, \tau) = (\sin \tau) e^{-2s}$$
$$= \sin\left(\frac{x}{t}\right) e^{-2(\ln t)}$$
$$= \frac{1}{t^2} \sin\left(\frac{x}{t}\right)$$

---

3. Solve the following convection problem in higher dimensions where a, b, c, d are given constants.

PDE: $au_x + bu_y + cu_t + du = 0 \quad -\infty < x < \infty, \; -\infty < y < \infty, \; 0 < t < \infty$

IC: $u(x, y, 0) = e^{-(x^2 + y^2)} \quad -\infty < x < \infty, \; -\infty < y < \infty$

Solution: Using the method of characteristics we let

$$\frac{dx}{ds} = a, \quad \frac{dy}{ds} = b, \quad \frac{dt}{ds} = c$$

which yield

Lesson 27: First-Order Equations (Method of Characteristics)

$$x(s) = as + c_1$$
$$y(s) = bs + c_2$$
$$t(s) = s + c_3$$

Using IC

$$x(0) = \tau_1,\ y(0) = \tau_2,\ t(0) = 0$$

we get $c_1 = \tau_1,\ c_2 = \tau_2,\ c_3 = 0$ and so

$$x = as + \tau_1$$
$$y = bs + \tau_2$$
$$t = s$$

Now, we can write the PDE and IC in terms of s and τ as

$$\text{ODE:}\quad \frac{du}{ds} + du = 0 \qquad 0 < s < \infty$$

$$\text{IC:}\quad u(\tau, 0) = e^{-(\tau_1^2 + \tau_2^2)}$$

which has the solution

$$u(s, \tau) = \sin \tau$$

or in terms of the original coordinates

$$u(x, t) = e^{-(\tau_1^2 + \tau_2^2)}\, e^{-ds}$$

now using the transformations

$$\tau_1 = x - at$$
$$\tau_2 = y - bt$$
$$s = t$$

we have

$$u(x,t) = e^{-(\tau_1^2 + \tau_2^2)}\, e^{-ds} = e^{-\left[(x-at)^2 + (y-bt)^2\right]}\, e^{-dt}$$

Lesson 27: First-Order Equations (Method of Characteristics)

4. Solve the following convection problem. Check your answer.

$$\text{PDE: } u_x + u_t + tu = 0 \quad -\infty < x < \infty, \ 0 < t < \infty$$

$$\text{IC: } u(x,0) = F(x) \quad -\infty < x < \infty$$

Solution: Using the method of characteristics we let

$$\frac{dx}{ds} = 1, \quad \frac{dt}{ds} = 1$$

which yield

$$x(s) = s + c_1$$
$$t(s) = s + c_2$$

Using IC

$$x(0) = \tau, \ t(0) = 0$$

we get $c_1 = \tau$, $c_2 = 0$ and so

$$x = s + \tau$$
$$t = s$$

If we eliminate s from the previous equations, we get the characteristic curves in the xt plane as a function of parameter τ, which are

$$x - t = \tau \quad (-\infty < \tau < \infty)$$

which is a family of straight lines in the xt.

Now, we can write the PDE and IC in terms of s and τ as

$$\text{ODE: } \frac{du}{ds} + su = 0 \quad 0 < s < \infty$$

$$\text{IC: } u(\tau, 0) = F(\tau)$$

The differential equation can be solved by separation of variables, writing

Lesson 27: First-Order Equations (Method of Characteristics)

$$\frac{du}{u} = -s\,ds \Rightarrow \ln u = -\frac{1}{2}s^2 + c \Rightarrow u = e^c e^{-s^2/2} = Ce^{-s^2/2}$$

where C is an arbitrary (positive) constant. Plugging in the IC gives $C = F(\tau)$ and so the solution of the ODE and IC is

$$u(\tau, s) = F(\tau) e^{-s^2/2}$$

or in terms of the original coordinates

$$u(x,t) = F(x-t) e^{-t^2/2}$$

---

5. It is possible to specify the solution u on a curve other than the usual initial time line $t = 0$. In fact, the differential equation doesn't have to involve the time variable at all (maybe u depends only on space variables). Solve the general first-order equation where $F(x,y)$ is a given function.

$$\text{PDE: } u_x + 2u_y + 2u = 0 \quad -\infty < x < \infty,\ -\infty < y < \infty$$

$$\text{IC: } u(x,y) = F(x,y) \text{ on the curve } C: y = x$$

Solution: Using the method of characteristics we let

$$\frac{dx}{ds} = 1, \quad \frac{dy}{ds} = 2$$

which yield

$$x(s) = s + c_1$$
$$y(s) = 2s + c_2$$

Since the IC is along the curve $y = x$, the IC for our equation in s is

$$x(0) = \tau,\ y(0) = \tau$$

from which we get $c_1 = \tau$, $c_2 = \tau$, hence

Lesson 27: First-Order Equations (Method of Characteristics)

$$x = s + \tau$$
$$y = 2s + \tau$$

Solving these equations for the new variables s and τ, we have

$$s = y - x$$
$$\tau = x - s = x - (y - x) = 2x - y$$

which, for each value of τ, gives characteristic curves in the xy plane as the family of straight lines with slope $2x$.

Now, we can write the original PDE and IC in terms of s and τ as

ODE: $\quad \dfrac{du}{ds} + 2u = 0 \quad -\infty < s < \infty$

IC: $\quad u(\tau, \tau) = F(\tau, \tau)$

which has the solution

$$u(s, \tau) = F(\tau, \tau) e^{-2s}$$

and so plugging in the values of s, τ in terms of x, y yields

$$u(x, y) = F(2x - y, 2x - y) e^{-2(y - x)}$$

Note that this function satisfies the IC and PDE. For example, if the initial condition were

$$F(x, y) = x^2 + y^2 \text{ on the line } y = x, \text{ then the solution is}$$

$$u(x, y) = \left[(2x - y)^2 + (2x - y)^2 \right] e^{-2(y - x)}$$
$$= 2(2x - y)^2 e^{-2(y - x)} \quad -\infty < x < \infty, -\infty < y < \infty$$

ΣΥΩZΞΠ

Lesson 28: Nonlinear First-Order Equations (Conservative Equations)

> 1. Solve the following IVP. Draw the solution for different values of time. What is your interpretation of the solution? What is the relationship between the flux and density in the problem?
>
> PDE: $u_t + u u_x = 0 \quad -\infty < x < \infty, \ -\infty < y < \infty$
>
> IC: $u(x,0) = \begin{cases} 0 & x < 0 \\ x & x \geq 0 \end{cases}$

Solution: We first find the characteristic curves (curves where u is a constant) starting at an arbitrary point $(x_0, 0)$, which are

$$x = x_0 + g(u(x_0, 0))t$$
$$= x_0 + u(x_0, 0)t$$
$$= \begin{cases} x_0 & x_0 < 0 \\ x_0 + x_0 t & x_0 \geq 0 \end{cases}$$

or solving for t when $x_0 \geq 0$ gives

$$t = \frac{1}{x_0}(x - x_0) \quad x_0 \geq 0$$

which is the family of straight lines drawn in Figure 26.1.

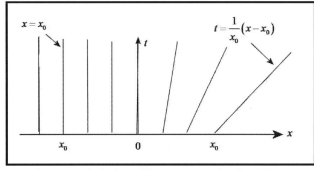

Figure 26.1 Characteristic lines

Lesson 28: Nonlinear First-Order Equations

Here $f_x = uu_x$ and so the flux $f(u)$ as a function of density u is

$$f(u) = \frac{1}{2}u^2.$$

---

2. Solve the following IVP. What is the flux-density relationship? Would you expect the solution to behave as it does? Compare the solution to the solution in Problem 1.

$$\text{PDE: } u_t + u^2 u_x = 0 \quad -\infty < x < \infty, \; 0 < t < \infty$$

$$\text{IC: } u(x,0) = \begin{cases} 0 & x < 0 \\ x & x \geq 0 \end{cases}$$

Solution: The implicit solution is

$$u = \varphi(x - g(u)t)$$

where

$$\phi(x) = \begin{cases} 0 & x < 0 \\ x & x \geq 0 \end{cases}$$

$$g(u) = u^2$$

Hence, for $x - u^2 t < 0$ or $u^2 > x/t$ we have $u(x,t) = 0$, and for $x - u^2 t \geq 0$ or $u^2 \leq x/t$ we have the implicit solution

$$u = \phi(x - u^2 t) = x - u^2 t$$

We can also find the characteristic curves (curves where u is a constant) starting at an arbitrary point $(x_0, 0)$, which are

$$x = x_0 + g(u(x_0,0))t$$
$$= x_0 + u^2(x_0,0)t$$
$$= \begin{cases} x_0 & x_0 < 0 \\ x_0 + x_0^2 t & x_0 \geq 0 \end{cases}$$

which is the family of straight lines drawn in Figure 26.2.

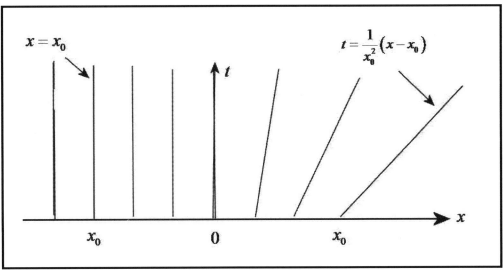

Figure 26.2 Characteristic lines

> 3. Suppose a non viscous liquid is traveling through a pipe and suppose the liquid leaks through the walls of the pipe according to the law $F(u) = ku$ (g/cm sec) where u is the density of the liquid, the flux of the liquid is $F(u) = u$, and the initial density of the liquid is $u(x,0) = \phi(x)$. What is the density $u(x,t)$ of the liquid for any x,t? What is your interpretation of the solution?

Solution: The (non) conservation problem is

$$u_t + f_x = -F(u)$$

But

$$f_x = \frac{df}{du} u_x = u_x$$

and so the problem becomes

Lesson 28: Nonlinear First-Order Equations

PDE: $u_t + u_x + ku = 0 \quad -\infty < x < \infty, \; 0 < t < \infty$

IC: $u(x,0) = \phi(x) \quad -\infty < x < \infty$

But this problem is linear so we use the method of characteristics to convert the PDE into an ODE along the characteristics. We do this by letting

$$\frac{dt}{ds} = 1 \Rightarrow t = s + c_1$$

$$\frac{dx}{ds} = 1 \Rightarrow x = s + c_2$$

Letting ICs $x(0) = \tau, \; t(0) = 0$, we find $c_1 = 0, \; c_2 = \tau$ and so $t = s, \; x = s + \tau$ (or $s = t, \; \tau = x - t$). Hence, we now have the ODE along the characteristics

$$\frac{du}{ds} + ku = 0$$

$$u(0) = \phi(\tau)$$

which has the solution

$$u(s, \tau) = \phi(\tau) e^{-ks}$$

or in terms of x, t

$$u(x,t) = \phi(x-t) e^{-kt}$$

The solution $u(x,t)$ moves to the right with constant velocity 1 and damps exponentially with damping rate $k > 0$.

---

4. What is the solution of the nonconservation equation if the loss of the liquid to the outside is $F(x) = 1/x$, the flux of the liquid is $f(u) = ku$, and initial density of the liquid is $u(x,0) = \varphi(x)$. Does your solution check? Does it make sense physically?

Solution: The IVP is

$$u_t + u_x + \frac{1}{x} = 0 \quad -\infty < x < \infty, \; 0 < t < \infty$$

$$u(x,0) = \phi(x) \quad -\infty < x < \infty$$

But this problem is linear so we use the method of characteristics to convert the PDE into an ODE along the characteristics. We do this by letting

$$\frac{dt}{ds} = 1 \Rightarrow t = s + c_1$$

$$\frac{dx}{ds} = 1 \Rightarrow x = s + c_2$$

Letting ICs $x(0) = \tau$, $t(0) = 0$, we find $c_1 = 0$, $c_2 = \tau$ and so $t = s$, $x = s + \tau$ (or $s = t$, $\tau = x - t$). Hence, we now have the ODE along the characteristics

$$\frac{du}{ds} + \frac{1}{s + \tau} = 0$$

$$u(0) = \phi(\tau)$$

The ODE has the general solution

$$u(s, \tau) = -\ln|s + \tau| + c$$

From the IC

$$u(0, \tau) = -\ln|\tau| + c = \phi(\tau)$$

we get

$$c = \phi(\tau) + \ln(\tau)$$

hence

$$u(s, \tau) = -\ln|s + \tau| + \phi(\tau) + \ln(\tau)$$

$$= \varphi(\tau) + \ln\left|\frac{\tau}{s + \tau}\right|$$

or in terms of x, t, we have

Lesson 28: Nonlinear First-Order Equations

$$u(x,t) = \phi(x-t) + \ln\left|\frac{x-t}{x}\right|$$

---

5. Verify that

$$u(x,t) = \phi[x - g(u)t]$$

is an implicit solution of the nonlinear IVP.

    PDE: $u_t + g(u)u_x = 0 \quad -\infty < x < \infty,\ 0 < t < \infty$

    IC: $\ \ u(x,0) = \phi(x) \quad -\infty < x < \infty$

Solution: Given

$$u = \phi[x - g(u)t]$$

differentiate each side of the equation with respect to t and x getting

$$u_t = \phi'(x - g(u)t)(-g(u) - g_u u_t t)$$
$$u_x = \phi'(x - g(u)t)(1 - g_u u_x t)$$

and substituting into

$$u_t + g(u)u_x$$

we get

$$u_t + g(u)u_x = \phi'(x - g(u)t)(-g(u) - g_u u_t t) + g(u)\left[\phi'(x - g(u)t)(1 - g_u u_x t)\right]$$
$$= -\phi'(x - g(u)t)\left[(u_t + gu_x)g_u t\right]$$

In order that this equation hold, it must be true that

$$u_t + g(u)u_x = 0$$

ΣΠNTZP

Lesson 29: Systems of PDEs

> 1. Write the system of PDEs in the following equations
>
> $$\frac{\partial u_1}{\partial t} = u_2$$
>
> $$\frac{\partial u_2}{\partial t} = u_3$$
>
> $$\frac{\partial u_3}{\partial t} = c^2 \frac{\partial u_2}{\partial x} + au_3 + bu_1$$
>
> in the form
>
> $$Au_t + Bu_x + Cu = 0$$
>
> where A, B and C are 3×3 matrices.

Solution:

$$A = \begin{bmatrix} 0 & 0 & 0 \\ 1 & 0 & 0 \\ 0 & 0 & 1 \end{bmatrix}, \quad B = \begin{bmatrix} 1 & 0 & 0 \\ 0 & -c^2 & 0 \\ 0 & 0 & 0 \end{bmatrix}, \quad C = \begin{bmatrix} 0 & -1 & 0 \\ 0 & 0 & -1 \\ -b & 0 & -a \end{bmatrix}$$

> 2. Find the eigenvalues and eigenvectors of the matrix
>
> $$A = \begin{bmatrix} 1 & 1 \\ 4 & 1 \end{bmatrix}$$

Solution: The eigenvalues are the roots of

$$|A - \lambda I| = \begin{vmatrix} 1-\lambda & 1 \\ 4 & 1-\lambda \end{vmatrix} =$$

$$= (1-\lambda)^2 - 4$$

$$= \lambda^2 - 2\lambda - 3$$

$$= (\lambda - 3)(\lambda + 1) = 0$$

Lesson 29: Systems of PDEs 185

Hence, the eigenvalues are $\lambda_1 = 3$, . To find the eigenvector

$$\vec{X}_1 = \begin{bmatrix} x_1 \\ x_2 \end{bmatrix}$$

associated with $\lambda_1 = 3$, we find all nonzero solutions of

$$\begin{bmatrix} 1 & 1 \\ 4 & 1 \end{bmatrix} \begin{bmatrix} x_1 \\ x_2 \end{bmatrix} = 3 \begin{bmatrix} x_1 \\ x_2 \end{bmatrix}$$

which gives

$$\vec{X}_1 = \begin{bmatrix} x_1 \\ x_2 \end{bmatrix} = c \begin{bmatrix} 1 \\ 2 \end{bmatrix}$$

Doing the same thing with the eigenvalue $\lambda_2 = -1$ we get the eigenvector $\vec{X}_2$ associated with $\lambda_2 = -1$, which is

$$\vec{X}_2 = \begin{bmatrix} 1 \\ -2 \end{bmatrix}$$

---

3. Using the eigenvalues and eigenvectors of Problem 2, find the general solution of the following system. Hint: First write the system in matrix form $u_t + Au = 0$.

$$\frac{\partial u_1}{\partial t} + \frac{\partial u_1}{\partial x} + \frac{\partial u_2}{\partial x} = 0$$

$$\frac{\partial u_2}{\partial t} + 4\frac{\partial u_1}{\partial x} + \frac{\partial u_2}{\partial x} = 0$$

Solution: Writing the system in matrix form $u_t + Au_x = 0$, we have

$$\begin{bmatrix} \dfrac{\partial u_1}{\partial t} \\ \dfrac{\partial u_2}{\partial t} \end{bmatrix} + \begin{bmatrix} 1 & 1 \\ 4 & 1 \end{bmatrix} \begin{bmatrix} \dfrac{\partial u_1}{\partial x} \\ \dfrac{\partial u_2}{\partial x} \end{bmatrix} = \begin{bmatrix} 0 \\ 0 \end{bmatrix}$$

In Problem 2 we found the eigenvectors $\vec{X}_1, \vec{X}_2$ of $\lambda_1 = 3, \lambda_2 = -1$ respectively and so we transform u to new dependent variable v via the transformation

$$\begin{bmatrix} u_1 \\ u_2 \end{bmatrix} = P\vec{v} = \begin{bmatrix} \vec{X}_1 : \vec{X}_2 \end{bmatrix} \begin{bmatrix} v_1 \\ v_2 \end{bmatrix} = \begin{bmatrix} 1 & 1 \\ 2 & -2 \end{bmatrix} \begin{bmatrix} v_1 \\ v_2 \end{bmatrix}$$

which transforms the original system

$$u_t + A u_x = 0$$

to the uncoupled system

$$v_t + \Lambda v_x = 0$$

which is

$$\begin{bmatrix} v_1 \\ v_2 \end{bmatrix} = \begin{bmatrix} 3 & 0 \\ 0 & -1 \end{bmatrix} \begin{bmatrix} v_1 \\ v_2 \end{bmatrix}$$

Written out these equation and their solutions are

$$\frac{\partial v_1}{\partial t} + 3 \frac{\partial v_1}{\partial x} = 0 \;\Rightarrow\; v_1(x,t) = \phi(x - 3t)$$

$$\frac{\partial v_2}{\partial t} - \frac{\partial v_2}{\partial x} = 0 \;\Rightarrow\; v_2(x,t) = \phi(x + t)$$

We can now find the solution vector u of the original system by plugging the vector v back in the equations $u = Pv$, or

$$\begin{bmatrix} u_1 \\ u_2 \end{bmatrix} = \begin{bmatrix} 1 & 1 \\ 2 & -2 \end{bmatrix} \begin{bmatrix} v_1 \\ v_2 \end{bmatrix}$$

$$= \begin{bmatrix} 1 & 1 \\ 2 & -2 \end{bmatrix} \begin{bmatrix} \phi(x-3t) \\ \phi(x+t) \end{bmatrix}$$

$$= \begin{bmatrix} \phi(x-3t) + \phi(x+t) \\ 2\phi(x-3t) - 2\phi(x+t) \end{bmatrix}$$

or in terms of the two components of the solution

Lesson 29: Systems of PDEs

$$u_1(x,t) = \phi(x-3t) + \phi(x+t)$$
$$u_2(x,t) = 2\phi(x-3t) - 2\phi(x+t)$$

---

4. Given the 2×2 matrix equation

$$u = Pv$$

defined in Problem 3, write the equation in scalar form to verify

$$\frac{\partial u}{\partial t} = P \frac{\partial v}{\partial t}$$

$$\frac{\partial u}{\partial x} = P \frac{\partial v}{\partial x}$$

Solution: We saw in Problem 3 that

$$P = \begin{bmatrix} 1 & 1 \\ 2 & -2 \end{bmatrix}$$

and so the system $u = Pv$ can be written as

$$\begin{bmatrix} u_1 \\ u_2 \end{bmatrix} = \begin{bmatrix} 1 & 1 \\ 2 & -2 \end{bmatrix} \begin{bmatrix} v_1 \\ v_2 \end{bmatrix}$$

or in scalar form

$$u_1 = v_1 + v_2$$
$$u_2 = 2v_1 - 2v_2$$

and if we differentiate these equations with respect to t we have

$$\frac{\partial u_1}{\partial t} = \frac{\partial v_1}{\partial t} + \frac{\partial v_2}{\partial t}$$

$$\frac{\partial u_2}{\partial t} = 2\frac{\partial v_1}{\partial t} - 2\frac{\partial v_2}{\partial t}$$

or in matrix form

$$\begin{bmatrix} \dfrac{\partial u_1}{\partial t} \\ \dfrac{\partial u_2}{\partial t} \end{bmatrix} = \begin{bmatrix} 1 & 1 \\ 2 & -2 \end{bmatrix} \begin{bmatrix} \dfrac{\partial v_1}{\partial t} \\ \dfrac{\partial v_2}{\partial t} \end{bmatrix}$$

or

$$\frac{\partial u}{\partial t} = P \frac{\partial v}{\partial t}$$

To prove the second equation

$$\frac{\partial u}{\partial x} = P \frac{\partial v}{\partial x}$$

simply replace t by x in the previous proof.

---

5. Show that the functions

$$u_1(x,t) = 2\phi(x-4t) - 2\psi(x+4t)$$
$$u_2(x,t) = \phi(x-4t) + \psi(x+4t)$$

satisfy the PDEs

$$\frac{\partial u_1}{\partial t} + 8 \frac{\partial u_2}{\partial x} = 0$$
$$\frac{\partial u_2}{\partial t} + 2 \frac{\partial u_1}{\partial x} = 0$$

Solution: Differentiating, we find

Lesson 29: Systems of PDEs

$$\frac{\partial u_1}{\partial t} = 2\phi'(x-4t)(-4) - 2\psi'(x+4t)(4) = -8\phi'(x-4t) - 8\psi'(x+4t)$$

$$\frac{\partial u_1}{\partial x} = 2\phi'(x-4t)(1) - 2\psi'(x+4t)(1) = 2\phi'(x-4t) - 2\psi'(x+4t)$$

$$\frac{\partial u_2}{\partial t} = \phi'(x-4t)(-4) + \psi'(x+4t)(4) = -4\phi'(x-4t) + 4\psi'(x+4t)$$

$$\frac{\partial u_2}{\partial x} = \phi'(x-4t)(1) + \psi'(x+4t)(1) = \phi'(x-4t) + \psi'(x+4t)$$

Careful examination of these equations verifies the desired PDEs.

ΡΤΞΩΤΘ

Lesson 30: The Vibrating Drumhead (Wave Equation in Polar Coordinates)

1. Substituting $u(r,\theta,t) = U(r,\theta)T(t)$ into the wave equation

$$u_t = c^2\left(u_{rr} + \frac{1}{r}u_r + \frac{1}{r^2}u_{\theta\theta}\right)$$

in order to separate it into the two equations

$$U_{rr} + \frac{1}{r}U_r + \frac{1}{r^2}U_{\theta\theta} + \lambda^2 U = 0$$

$$T' + \lambda^2 c^2 T = 0$$

Solution: Plugging

$$u(r,\theta,t) = U(r,\theta)T(t)$$

into

$$u_{tt} = c^2\left(u_{rr} + \frac{1}{r}u_r + \frac{1}{r_2}u_{\theta\theta}\right)$$

we get

$$UT'' = c^2\left(U_{rr}T + \frac{1}{r}U_r T + \frac{1}{r_2}U_{\theta\theta}T\right)$$

Dividing by $U(r,\theta)T(t)$ gives

$$\frac{T''}{c^2 T} = \frac{U_{rr}}{U} + \frac{1}{r}\frac{U_r}{U} + \frac{1}{r^2}U_{\theta\theta}$$

Setting each side of this equation to a constant, say $-\lambda^2$ (for convenience), we have the desired result

Lesson 30: The Vibrating Drumhead

$$U_{rr} + \frac{1}{r}U_r + \frac{1}{r^2}U_{\theta\theta} + \lambda^2 U = 0$$

$$T' + \lambda^2 c^2 T = 0$$

---

2. Substitute $U(r,\theta) = R(r)\Theta(\theta)$ into the following Helmoltz BVP. Why do we choose the separation constant to be $0,1,2,..$ which we call n^2.

PDE: $U_{rr} + \frac{1}{r}U_r + \frac{1}{r^2}U_{\theta\theta} + \lambda^2 U = 0$

BC: $U(1,\theta) = 0$

to get

BVP: $\begin{cases} r^2 \dfrac{d^2 R}{dr^2} + r \dfrac{dR}{dr} + (\lambda^2 r^2 - n^2) R = 0 \\ R(1) = 0 \end{cases}$

ODE: $\dfrac{d^2\Theta}{d\theta^2} + n^2 \Theta = 0$

Solution: Direct substitution (similar to Problem 1).

---

3. Solve

PDE: $u_{tt} = u_{rr} + \dfrac{1}{r}u_r + \dfrac{1}{r^2}u_{\theta\theta}$ $0 < r < 1,\ 0 < \theta < 2\pi,\ 0 < t < \infty$

BC: $u(1,\theta,t) = 0$ $0 < \theta < 2\pi,\ 0 < t < \infty$,

IC: $\begin{cases} u(r,\theta,0) = 1 - r^2 \\ u_t(r,\theta,0) = 0 \end{cases}$ $0 \le r < 1,\ 0 \le \theta < 2\pi$

Solution: Since the BC and ICs do not depend on θ, then neither will the solution, hence we seek a solution of the form

$$u(r,t) = R(r)T(t)$$

Plugging this in the PDE (neglecting the term $u_{\theta\theta}$), we get

$$T''R = TR'' + \frac{1}{r}TR'$$

and dividing by RT gives

$$\frac{T''}{T} = \frac{R''}{R} + \frac{1}{r}\frac{R'}{R}$$

Setting each side of this equation to $-\lambda^2$ we arrive the ODE

$$T'' + \lambda^2 T = 0$$

and an eigenvalue/eigenvector problem

ODE: $r^2 R'' + rR' + \lambda^2 r^2 R = 0, \ 0 < r < 1$

BC: $\begin{cases} R(1) = 0 \\ R(r) < \infty \end{cases}$

This problem is called a Sturm-Liouville problem and its solutions are called eigenvalues and the possible values of λ are called the eigenvalues of the Sturm-Liouville system. In this problem there are an infinite number of functions that satisfy both the ODE and BCs, and they are

$$R_m(r) = J_0(\lambda_{0m} r), \ m = 0, 1, 2, \ldots,$$

where the first four eigenvalues λ_{0m} are

$$\lambda_{01} \doteq 2.40$$
$$\lambda_{02} \doteq 5.52$$
$$\lambda_{03} \doteq 8.65$$
$$\lambda_{04} \doteq 11.79$$
$$\ldots \ \ldots$$

and the corresponding eigenfunctions.

Lesson 30: The Vibrating Drumhead   193

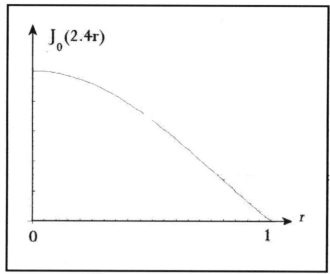

Figure 30.1a) First eigenfunction $J_0(2.4r)$

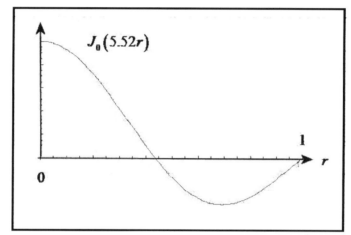

Figure 30.1b) Second eigenfunction $J_0(5.2r)$

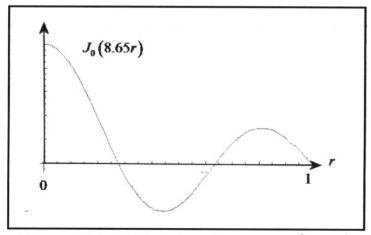

Figure 30.1c) Third eigenfunction $J_0(8.65r)$

We now solve the ODE
$$T'' + \lambda^2 T = 0$$

getting
$$T_n(t) = \cos(\lambda_{0n} t)$$

hence we have a sum of product solutions solutions

$$u(r,t) = \sum_{m=1}^{\infty} a_m J_0(k_{0m} r) \cos(k_{0m} t)$$

where the coefficients a_n are determined so $u(r,t)$ satisfies the ICs

$$u(r,0) = \sum_{n=1}^{\infty} a_n J_0(k_{0n} r) = 1 - r^2 \quad 0 \leq r \leq 1, \ 0 \leq \theta < 2\pi$$

We find the coefficients by multiplying each side of the equation by $rJ_0(k_{0m} r)$ and integrating each side of the equation with respect to r on the interval $[0,1]$, and using the orthogonality of the eigenfunctions

$$\int_0^1 r J_0(k_{0m} r) J_0(k_{0n} r) dr = \begin{cases} 0 & m \neq n \\ \dfrac{1}{2} J_1(k_{0n}) & m = n \end{cases}$$

(where $J_1(r)$ is the first-order Bessel function, we get

$$\int_0^1 r(1 - r^2) J_0(k_{0m} r) dr = a_m \int_0^1 J_0^2(k_{0m} r) dr = \frac{1}{2} a_m J_1^2(k_{0m})$$

$$a_m = \frac{2 \int_0^1 r(1 - r^2) J_0(k_{0m} r) dr}{J_1^2(k_{0m})}$$

The integrals for a_n can be evaluated on a computer. The Mathematica statement for finding a_1 is

Integrate$\left[2*r*(1-r*r)*\text{Bessel}[0, 2.4r], \{r, 0, 1\} \right]$ / Bessel$[1, 2.4]\wedge 2$

and gives the values $a_1 \doteq 1.1$, $a_2 \doteq -0.34$, $a_3 \doteq 0.045$. After finding the coefficients (or 3 or 4) the solution is then

$$u(r,t) = \sum_{m=1}^{\infty} a_m J_0(k_{0m}r) \cos(k_{0m}t)$$
$$\doteq 1.1 J_0(2.4r)\cos(2.4t) - 0.34 J_0(5.52r)\cos(5.52t) + 0.045 J_0(8.65r)\cos(8.65t)$$

4. Solve

PDE: $u_{tt} = \nabla^2 u \quad 0 < r < 1, \; 0 < \theta < 2\pi, \; 0 < t < \infty$

BC: $u(1, \theta, t) = 0 \quad 0 \leq \theta < 2\pi, \; 0 < t < \infty,$

IC: $\begin{cases} u(r, \theta, 0) = J_0(2.4r) \\ u_t(r, \theta, 0) = 0 \end{cases} \quad 0 \leq r \leq 1, \; 0 \leq \theta < 2\pi$

Solution:
$$u(r,t) = J_0(2.4r)\cos(2.4t)$$

5. Solve the following IBVP. What is the highest frequency of vibration?

PDE: $u_{tt} = \nabla^2 u \quad 0 < r < 1, \; 0 < \theta < 2\pi, \; 0 < t < \infty$

BC: $u(1, \theta, t) = 0 \quad 0 < \theta < 2\pi, \; 0 < t < \infty,$

IC: $\begin{cases} u(r, \theta, 0) = J_0(2.4r) - \dfrac{1}{2} J_0(8.65r) + \dfrac{1}{4} J_0(14.93r) \\ u_t(r, \theta, 0) = 0 \end{cases}$

Solution: The general solution of the vibrating circular drumhead that is independent of θ is

$$u(r,t) = \sum_{m=1}^{\infty} a_m J_0(k_{0m}r) \cos(k_{0m}t)$$

The general idea is that we expand the IC $u(r,\theta,0) = f(r)$ as

$$g(r) = a_1 J_0(k_{01}r) + a_2 J_0(k_{02}r) + a_3 J_0(k_{03}r) + \cdots$$

then simply add the factor $\cos(k_{0n}t)$ in each term, getting the solution

$$u(r,t) = a_1 J_0(k_{01}r)\cos(k_{01}t) + a_2 J_0(k_{02}r)\cos(k_{02}t) + a_3 J_0(k_{03}r)\cos(k_{03}t) + \cdots$$

In this problem the IC is *already* expanded as a series of Bessel functions

$$u(r,\theta,0) = J_0(2.4r) - \frac{1}{2} J_0(8.65r) + \frac{1}{4} J_0(14.93r)$$

and so the solution is simply

$$u(r,\theta) = J_0(2.4r)\cos(2.4t) - \frac{1}{2} J_0(8.65r)\cos(8.65t)$$
$$+ \frac{1}{4} J_0(14.93r)\cos(14.93t)$$

---

6. Graph the following Bessel functions.

   a) $J_0(5.52r)$

   b) $J_0(14.93r)$

Solution: The following graphs were done on Mathematia. Instructions for graphing the functions are given below the graphs. You could also graph them by going to the webpage www.wolframalpha.com and entering the commands under the graphs.

Lesson 30: The Vibrating Drumhead

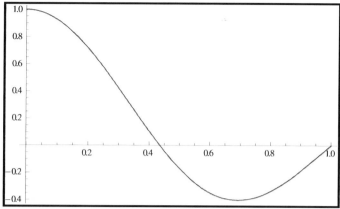

Figure 26.3 Plot[BesselJ[0,5.52r],{r,0,1}]

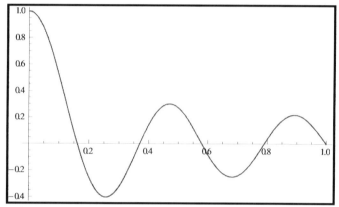

Figure 26.4 Plot[BesselJ[0,14.93r],{r,0,1}]

ΣΡΠΘΟΖ

Professor Pickle and the Omicron Affair

When the legendary Professor Pickle touched down at Duckworth Academy for boys, it was rumored that a conspiracy of ravens landed on the Commons. The Clock on the Old Tower chimed six bells. The ivy at the north end of Mangrove Hall turned brown and fell from the vines. And not least of all, students of mathematics prayed for divine intervention as all hope of horseplay in their upcoming calculus 1 class fled like vermin on a sinking ship.

During my first year at Duckworth, I was one of the unfortunate souls assigned to Professor Pickle's beginning calculus class. I still recall the dire expressions on the faces of my fellow greenhorns who were internally bawling and thinking to themselves, "All hope abandon ye who enter here."

My old friend Crowder was so scared he was shaking like a wounded gazelle. He knew he was digging his own grave, and true to his word, he never made it through.

Although the vast majority of students in the class floundered in the wake of Pickle's endless barrage of arcane symbols, he was a mathematical utopia for the mathematical aristocracy, who were on speaking terms with Pickle's byzantine vocabulary, mopping up his every premise and inference. Professor Pickle was a mathematical prophet spreading the Word to fertile minds.

"Ah, mathematics, it's the poetry of logical ideas," they cheered. "It's the most beautiful creation of the human spirit."

For us hoi polloi drowning in an ocean of unintelligible babble, the human spirit was on holiday.

It generally took Pickle ten minutes to fill all the panels of the blackboard with equations, after which he returned to the starting line and began refilling them with replacements. I wrote like the wind, but could never begin to keep ahead of him as he'd come up behind me, lap me, and disappear down the board.

Then, all of a sudden and a thousand equations later, he would stop abruptly, stand back from the board, point to a rather odious-looking equation, and ask himself,

"Hmmmmmmmm, now let's see what we have here?"

By now, the class was scattered all over the board taking copious notes, but upon hearing Pickle's hallmark query, everyone would abruptly freeze and look in a befuddled manner at Pickle's prize equation. After a few seconds, however, everyone's look of befuddlement changed to horror as they knew that Pickle's question wasn't always *rhetorical*, and there was the distinct possibility he might ask one of *them*, what do we have here?

"Well, mister ---- " was the dreaded phrase each member of the class feared as no one dared look him straight in the face. One could hear the shuffling of papers and the scratching of a pencil, like a rat scratching for grain, by some poor retch in a pathetic attempt to say why he shouldn't be interrupted from busy note-taking.

Then, after Pickle grilled some helpless soul within an inch of his academic life, he would reload his chalk and race off down the board, leaving the class wallowing in his wake. Then, ten minutes later and another million equations, as if by clockwork, he would stop again and ask,

"Hmmmmmmm, now let's see what we have here?"

If we hadn't a clue what was there before, I always wondered how he expected us to know what was there now? Of course, I wasn't about to raise the issue now, especially when I was so busy shuffling my papers.

I estimated that over the entire semester Pickle used his trademark phrase a thousand times. I actually saw the object of Pickle's famous question one time, giving me a semester's average of 0.001.

However, as much as Pickle is remembered at Duckworth for his Teutonic manner and famous utterance, he will always be remembered by what is known as the *Omicron Affair*.

The first thing you should know about Pickle was how he loved mathematical symbols. Generally somewhere in his first sentence would be the phrase

given an $\varepsilon > 0$ there exists a $\delta > 0$ such that ...

Before long the blackboard was filled with hundreds of equations that looked more or less like

$$\frac{\partial^2 \Psi}{\partial \theta^2} = \frac{1}{\varphi}\left\{\lambda + \frac{2^n \eta}{\zeta}\frac{\phi''}{\phi}\right\} + \frac{2\delta}{\Pi}\iiint_\Omega \Pi_\alpha(\sigma)d\sigma$$

During one of his more ambitious lectures, Pickle ran through his favorite Greek letters of α, β, γ and δ and so on, eventually getting to some of the lesser used letters like $\lambda, \theta,$ and ω. For the average math teacher, a few Greek letters suffice, maybe an occasional $\phi, \zeta,$ or μ when the discussion becomes a little verbose.

During this particular lecture, however, Pickle seemed to run through the entire Greek alphabet, but had five minutes of class time left. He paced back and forth in front of the board, mumbling to himself and scratching his head. It was the first time the class had ever seen Pickle stumped. A classmate carefully shot me a half-grin out the corner of his mouth.

Finally, Pickle lifted a finger in the air. "*Aha,*" he announced proudly. The entire class leaned forward, the first time actually awaiting the answer to one of his questions.

"*Omicron!*"

"Omicron?"

"*Omicron!*" Pickle repeated. "We haven't used *omicron!*"

My God, I thought to myself. No one ever uses omicron. You can't tell it from the letter 'o' or a zero. The man has completely devoured the Greek alphabet.

"We're seeing mathematical history being made here today," a classmate muttered out the side of his mouth.

Well, that's it. For those of you who have never taken a course in mathematics, you probably don't appreciate Pickle's accomplishment, but it just might be the only time in the history of mathematics that a mathematics teacher has used the entire Greek alphabet in a single lecture. I consider myself fortunate to have witnessed it firsthand.

I only wish I knew what it was all about[1].

---

[1] There is some truth to this story. The author experienced such an incident in graduate school in a differential equations class, the only difference being the professor's name was not Pickle.

Section 4: Elliptic Type Problems

Lesson 31: The Laplacian (an Intuitive Description)

> 1. What does the wave equation in spherical coordinates
> $$u_{tt} = c^2 \left[u_{rr} + \frac{2}{r} u_r + \frac{1}{r^2} u_{\phi\phi} + \frac{\cot\phi}{r^2} u_\phi + \frac{1}{r^2 \sin^2\phi} u_{\theta\theta} \right]$$
> reduce to if the solution u depends only on r and t?

Solution: The general wave equation in (both forms) spherical coordinates is
$$u_{tt} = c^2 \left[\frac{1}{r^2}(r^2 u_r)_r + \frac{1}{r^2 \sin\phi}(\sin\phi\, u_\phi)_\phi + \frac{1}{r^2 \sin^2\phi} u_{\theta\theta} \right]$$
$$= c^2 \left[u_{rr} + \frac{2}{r} u_r + \frac{1}{r^2} u_{\phi\phi} + \frac{\cot\phi}{r^2} u_\phi + \frac{1}{r^2 \sin^2\phi} u_{\theta\theta} \right]$$

Hence, if u depends only on t and r, all partial derivatives of u with respect to θ and ϕ are zero, and so the equation reduces to

$$u_{tt} = c^2 \left(u_{rr} + \frac{2}{r} u_r \right)$$

A good exercise is to work out the derivatives in the first above form of the wave equation to show that it is the same as the second form.

> 2. What does the wave equation $u_{tt} = c^2 \nabla^2 u$ in polar coordinates reduce to if the solution u only depends only on r and t?

Solution: The wave equation in polar coordinates is

$$u_{tt} = c^2 \left(u_{rr} + \frac{1}{r} u_r + \frac{1}{r^2} u_{\theta\theta} \right)$$

so if u depends only on t and r, we have $u_{\theta\theta} = 0$, and so the equation reduces to

$$u_{tt} = c^2 \left(u_{rr} + \frac{1}{r} u_r \right)$$

Note that this equation is *almost* the same as the spherical equation in Problem 1 that only depends on r (coefficient of $1/r$ in this problem for polar coordinates, compared with a coefficient of $2/r$ for spherical coordinates), but this small difference results in a *big* difference in the solution. The solution of the wave equation in spherical coordinates (3 dimensions) that only depends on r is $u(r) = 1/r$, whereas the solution of the wave equation in polar coordinates (2 dimensions) that only depends on r is $u(r) = \ln(1/r)$.

> 3. What is Laplace's equation $\nabla^2 u = 0$ in polar coordinates if the solution u only depends only on r? What are the solutions of this equation? These solutions are called the *circularly symmetric potentials* in two dimensions.

Solution: In general Laplace's equation in polar coordinates is

$$\nabla^2 u \equiv u_{rr} + \frac{1}{r} u_r + \frac{1}{r^2} u_{\theta\theta} = 0$$

If the function u only depends on r and is independent on θ, then we have $u_{\theta\theta} = 0$ and so the PDE becomes the ODE

$$\frac{d^2 u}{dr^2} + \frac{1}{r} \frac{du}{dr} = 0$$

Lesson 31: The Laplacian (an Intuitive Description)

which although is a second-order equation, it is a first-order equation in the derivative $U(t) = du/dr$, giving us the first-order ODE

$$\frac{dU}{dr} + \frac{1}{r}U = 0$$

Separating variables, we have

$$\frac{dU}{U} = \left(-\frac{1}{r}\right)dr$$

or

$$\ln U(r) = -\ln r + c = \ln\left(\frac{1}{r}\right) + c$$

Solving for $U(r)$ gives

$$U(r) = e^c e^{\ln(1/r)} = \frac{c}{r}$$

where we simply replace e^c by c (you can verify that c/r satisfies the ODE for any real constant c and not just positive constants e^c). Now, since $U(t) = du/dr$, we integrate to get the general solution

$$u(r) = c_1 + c_2 \ln r$$

where c_1 and c_2 are arbitrary constants. This solution is called the *logarithmic potential* and is often written in equivalent form

$$u(r) = c_1 + c_2 \ln\left(\frac{1}{r}\right)$$

since $\ln(1/r) = -\ln r$.

---

4. What is Laplace's equation $\nabla^2 u = 0$ in spherical coordinates if the solution u only depends only on r? What are the solutions of this equation? These solutions are called the *spherically symmetric potentials* in three dimensions

Lesson 31: The Laplacian (an Intuitive Description)

Solution: Laplace's equation in spherical coordinates can be expressed either in the more compact form (so-called divergence form)

$$\left(r^2 u_r\right)_r + \frac{1}{\sin\phi}\left(\sin\phi\, u_\phi\right)_\phi + \frac{1}{\sin^2\phi} u_{\theta\theta} = 0$$

or in the more expanded form

$$u_{rr} + \frac{2}{r} u_r + \frac{1}{r^2} u_{\phi\phi} + \frac{\cot\phi}{r^2} u_\phi + \frac{1}{r^2 \sin^2\phi} u_{\theta\theta} = 0$$

If the solution u only depends on r, the partial derivatives with respect to θ and ϕ are zero, and hence Laplace's equation reduces to the ODE

$$\frac{d^2 u}{dr^2} + \frac{2}{r}\frac{du}{dr} = 0$$

which can easily be solved by first letting $U(t) = du/dr$, after which we get

$$\frac{dU}{dr} + \frac{2}{r} U = 0$$

Separating variables, we have

$$\frac{dU}{U} = \left(-\frac{2}{r}\right) dr$$

or

$$\ln U = -2 \ln r + c = \ln r^{-2} + c = \ln\left(\frac{1}{r^2}\right) + c$$

Solving for $U(r)$, we get

$$U(r) = \frac{c}{r^2}$$

and since $U(t) = du/dr$ we integrate this equation to get

Lesson 31: The Laplacian (an Intuitive Description)

$$u(r) = c_1 + \frac{c_2}{r}$$

where c_1 and c_2 are arbitrary constants. Note the similarity between the Laplacian in 3 dimensions and the Laplacian in 2 dimensions that only depends on the distance r from the origin.

$$\text{2 dimensions} = \frac{d^2 u}{dr^2} + \frac{1}{r}\frac{du}{dr} = 0 \Rightarrow u(r) = c_1 + c_2 \ln\left(\frac{1}{r}\right)$$

$$\text{3 dimensions} = \frac{d^2 u}{dr^2} + \frac{2}{r}\frac{du}{dr} = 0 \Rightarrow u(r) = c_1 + c_2 \left(\frac{1}{r}\right)$$

---

5. What can you say about the (harmonic) surface $u(x,t) = xy$?

Solution: By quick examination, you will discover that $u(x,t) = xy$ satisfies $u_{xx} = u_{yy} = 0$ and hence satisfies Laplace's equation

$$u_{xx} + u_{yy} = 0$$

Such functions are called *harmonic functions*. Although there are many functions that satisfy the Laplace's equation, they have important properties in common, one being the 'mean value property' which states that if you take a sphere (or circle in 2 dimensions) surrounding a point, then the average value of u on the sphere (or circle) is the value of u at the center of the sphere (or circle). Harmonic functions are important in physics, engineering, ...

If you study complex variables, you will discover that the real u and complex v parts of an analytic function (i.e. differentiable) satisfy Laplace's equation. A few examples are

$$z^2 = (x+iy)^2 = \underbrace{x^2 - y^2}_{u} + i\underbrace{(2xy)}_{v}$$

$$e^{az} = e^{a(x+iy)} = e^{ax}e^{iay} = \underbrace{e^{ax}\cos ay}_{u} + i\underbrace{e^{ax}\sin ay}_{v}$$

$$\frac{1}{z^n} = \frac{1}{r^n}e^{-in\theta} = \underbrace{\frac{1}{r^n}\cos n\theta}_{u} - \underbrace{\frac{1}{r^n}\sin n\theta}_{v}, \quad z \neq 0$$

$$\ln z = \ln(re^{i\theta}) = \underbrace{\ln r}_{u} + i\underbrace{\theta}_{v}$$

6. Transform the two-dimensional Laplacian

$$\nabla^2 u = u_{xx} + u_{yy}$$

to the polar coordinate form

$$\nabla^2 u \equiv u_{rr} + \frac{1}{r}u_r + \frac{1}{r^2}u_{\theta\theta} = 0$$

by the transformation

$$x = r\cos\theta$$
$$y = r\sin\theta$$

Solution: To change Laplace's equation in Cartesian coordinates to Laplace's equation in polar coordinates, we make the transformation

$$x = r\cos\theta$$
$$y = r\sin\theta$$

The following diagram may be helpful in computing the partial derivatives required to make this transformation.

Lesson 31: The Laplacian (an Intuitive Description)

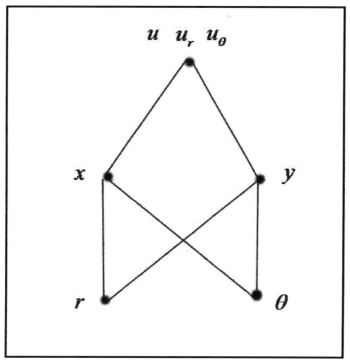

Figure 31.1 Diagram for computing partial derivatives of u, u_r and u_θ

First observe

$$x_r = \cos\theta, \ x_\theta = -r\sin\theta$$
$$y_r = \sin\theta, \ y_\theta = r\cos\theta$$

We begin by taking *first* partial derivatives

$$u_r = u_x x_r + u_y y_r = u_x(\cos\theta) + u_y(\sin\theta)$$
$$u_\theta = u_x x_\theta + u_y y_\theta = u_x(-r\sin\theta) + u_y(r\cos\theta)$$

We now continue on and compute the second derivatives, still using the diagram in Figure 31.1 and the fact that $u_{xy} = u_{yx}$. The second partial derivatives take a little more thought, but we have (watch carefully)

$$u_{rr} = \frac{\partial}{\partial r} u_r$$

$$= \frac{\partial}{\partial r}\left[u_x(\cos\theta) + u_y(\sin\theta)\right]$$

$$= \left[\left(\frac{\partial}{\partial r}u_x\right)\cos\theta + u_x\frac{\partial}{\partial r}\cos\theta\right] + \left[\left(\frac{\partial}{\partial r}u_y\right)\sin\theta + u_y\frac{\partial}{\partial r}\sin\theta\right]$$

but

$$\left(\frac{\partial}{\partial r}u_x\right)\cos\theta = u_{xx}\cos^2\theta + u_{xy}\sin\theta\cos\theta$$

$$\frac{\partial}{\partial r}\cos\theta = 0$$

$$\left(\frac{\partial}{\partial r}u_y\right)\sin\theta = u_{xy}\sin\theta\cos\theta + u_{yy}\sin^2\theta$$

$$\frac{\partial}{\partial r}\sin\theta = 0$$

and so

$$u_{rr} = u_{xx}\cos^2\theta + 2u_{xy}\sin\theta\cos\theta + u_{yy}\sin^2\theta$$

Also

Lesson 31: The Laplacian (an Intuitive Description)

$$u_{\theta\theta} = \frac{\partial}{\partial\theta}u_\theta$$

$$= \frac{\partial}{\partial\theta}\left[u_x(-r\sin\theta) + u_y(r\cos\theta)\right]$$

$$= \frac{\partial}{\partial\theta}\left[u_x(-r\sin\theta)\right] + \frac{\partial}{\partial\theta}\left[u_y(r\cos\theta)\right]$$

$$= \left[\left(\frac{\partial}{\partial\theta}u_x\right)(-r\sin\theta) + u_x\left(\frac{\partial}{\partial\theta}(-r\sin\theta)\right)\right] + \left[\left(\frac{\partial}{\partial\theta}u_y\right)(r\cos\theta) + u_y\left(\frac{\partial}{\partial\theta}(r\cos\theta)\right)\right]$$

$$= \left[u_{xx}(r^2\sin^2\theta) + u_{xy}(r^2\sin\theta\cos\theta) - u_x(r\cos\theta)\right]$$
$$+ \left[u_{xy}(-r^2\sin\theta\cos\theta) + u_{yy}(r^2\cos^2\theta)\right] - u_y(r\sin\theta)$$

Plugging $u_r, u_{rr}, u_{\theta\theta}$ in Laplace's equation in polar coordinates, we see

$$u_{rr} + \frac{1}{r}u_r + \frac{1}{r^2}u_{\theta\theta} = \left[u_{xx}\cos^2\theta + u_{yy}\sin^2\theta + 2u_{xy}(\sin\theta)(\cos\theta)\right] + \frac{1}{r}\left[u_x(\cos\theta) + u_y(\sin\theta)\right]$$

$$+ \frac{1}{r^2}\left[\begin{array}{l}\left[u_{xx}(r^2\sin^2\theta) + u_{xy}(r^2\sin\theta\cos\theta) - u_x(r\cos\theta)\right] \\ + \left[u_{xy}(-r^2\sin\theta\cos\theta) + u_{yy}(r^2\cos^2\theta)\right] - u_y(r\sin\theta)\end{array}\right]$$

$$= u_{xx} + u_{yy} \quad \text{(after considerable algebra)}$$

There is another way to do this problem. We could solve for r, θ in the transformation

$$x = r\cos\theta$$
$$y = r\sin\theta$$

getting

$$r = \sqrt{x^2 + y^2}$$
$$\theta = \tan^{-1}\left(\frac{y}{x}\right)$$

and then compute the derivatives u_{xx} and u_{yy} in terms of r, θ and plug them into $u_{xx} + u_{yy} = 0$ and arrive at Laplace's equation in polar coordinates; but this is more difficult since the partial derivatives of r and θ with respect to x and y are more messy.

Note: If you like computing complicated partial derivatives, try changing the three-dimensional Laplace's equation in Cartesian coordinates

$$u_{xx} + u_{yy} + u_{zz} = 0$$

to Laplace's equation in spherical Laplace's equation

$$u_{rr} + \frac{2}{r}u_r + \frac{1}{r^2}u_{\phi\phi} + \frac{\cot\phi}{r^2}u_\phi + \frac{1}{r^2 \sin^2\phi}u_{\theta\theta} = 0$$

by means of the change of variables

$$x = r\sin\phi\cos\theta$$
$$y = r\sin\phi\sin\theta$$
$$z = r\cos\phi$$

Real fun!

ΟΓΔΒΠΘ

Lesson 32: General Nature of Boundary Value Problems

> 1. Based on intuition, can you find the solution of the following Dirichlet problem.
> $$\text{PDE: } u_{rr} + \frac{1}{r}u_r + \frac{1}{r^2}u_{\theta\theta} = 0, \ 0 < r < 1$$
> $$\text{BC: } u(1,\theta) = \sin\theta \quad 0 \le \theta < 2\pi$$

Solution: If we imagine Laplace's equation as a crude model for a rubber sheet, imagine stretching the rubber sheet around the outside $(r=1)$ in the shape of $u(1,\theta) = \sin\theta$. If you make the observation that $\sin\theta = -\sin(\theta + \pi)$, then the rubber sheet is stretched in the opposite direction on the opposite side of the origin; i.e. 180 degrees apart. For that reason, one might be tempted to think the rubber sheet is linear in r and maybe the solution is

$$u(r,\theta) = r\sin\theta$$

This function clearly satisfies the BC, and if we plug it in the PDE, we find

$$u_{rr} + \frac{1}{r}u_r + \frac{1}{r^2}u_{\theta\theta} = 0 + \frac{1}{r}\sin\theta + \frac{1}{r^2}(-r\sin\theta) = 0$$

> 2. Does the following Neumann problem have a solution inside the circle?
> $$\text{PDE: } u_{rr} + \frac{1}{r}u_r + \frac{1}{r^2}u_{\theta\theta} = 0, \ 0 < r < 1$$
> $$\text{BC: } u_r(1,\theta) = \sin^2\theta \quad 0 \le \theta < 2\pi$$

Solution: The net flux across the circle must satisfy

$$\text{net flux across the circle} = \int_0^{2\pi} u_r(1,\theta)d\theta = 0$$

and in this problem we have

$$\text{net flux across the circle} = \int_0^{2\pi} \sin^2\theta\, d\theta > 0$$

and so there is no solution since the total flux (integral of the normal derivative on the boundary) across the boundary for the Neumann problem must be zero. The proof that the total flux is zero comes from Gauss' theorem in multivariable calculus that if R is a bounded region of $\Re^2$ (it also holds in three dimensions but we state the result in two dimensions) with boundary ∂R and if F is a vector field defined over V, then the integral of the divergence of F over R is the line integral of the normal component of F on the boundary curve ∂R. Mathematically stated

$$\iiint_R (\nabla \cdot F)\, dR = \oiint_{\partial R} (F \cdot n)\, dr$$

Now, if the vector field F is the gradient vector, i.e. $F = \nabla u$ and the region is a circle of radius 1, then $\nabla \cdot \nabla u = \nabla^2 u$ and Gauss' theorem reduces to

$$\int_0^{2\pi} \int_0^1 \left(u_{rr} + \frac{1}{r} u_r + \frac{1}{r^2} u_{\theta\theta} \right) dr\, d\theta = \int_0^{2\pi} u_r(1,\theta)\, d\theta$$

But, if Laplace's equation holds, the integrand on the left is zero, hence the flux integral on the right is zero.

---

3. Give an intuitive description of the solution of the following BV for different values of h.

PDE: $u_{rr} + \frac{1}{r} u_r + \frac{1}{r^2} u_{\theta\theta} = 0$, $0 < r < 1$

BC: $u_r(1,\theta) + h[u(1,\theta) - \sin\theta] = 0$ $0 \leq \theta < 2\pi$

Solution: When $h = 0$ the BC is the Neumann BC $u_r(1,\theta) = 0$. The larger h becomes the more weight is placed on making the BC look like the Dirichlet BC $u(1,\theta) = \sin\theta$. You can think of h as a parameter going from a Neumann problem to a Dirichlet problem. For 'medium sized' h, the BC

Lesson 32: General Nature of Boundary Value Problems

acts like a thermostat adding or subtracting heat depending on whether $u(1,\theta) > \sin\theta$ or $u(1,\theta) > \sin\theta$.

---

4. What is the boundary value problem one must solve to find the steady-state solution of

$$\text{PDE:} \quad u_{tt} = u_{xx} - u_t + u \quad 0 < x < 1, \ 0 < t < \infty$$

$$\text{BCs:} \quad \begin{cases} u(0,t) = 0 \\ u(1,t) = 0 \end{cases} \quad 0 < t < \infty$$

$$\text{ICs:} \quad \begin{cases} u(x,0) = \sin(3\pi x) \\ u_t(x.0) = 0 \end{cases} \quad 0 \le x < 1$$

Solution: We suspect there will be a steady state solution due to the damping term $-u_t$. Setting $u_t = u_{tt} = 0$ in the PDE and along with the BC, we get

$$\text{ODE:} \quad \frac{d^2u}{dx^2} + u = 0 \quad 0 < x < 1$$

$$\text{BC:} \quad \begin{cases} u(0) = 0 \\ u(1) = 0 \end{cases}$$

The general solution of the ODE is

$$u(x) = c_1 \sin x + c_2 \cos x$$

where c_1 and c_2 are arbitrary constants. Plugging this general solution in the BC gives

$$c_2 = 0$$
$$c_1 \sin(1) = 0 \Rightarrow c_1 = 0$$

Hence, the steady state is $u(x) = 0$, $0 \le x \le 1$. If you remember from your ODEs the y' term in the damped harmonic oscillator equation

$$y'' + y' + y = 0$$

Lesson 32: General Nature of Boundary Value Problems

is what causes the solutions to go to zero, the same principle in PDEs. The first time derivative in the somewhat similar PDE

$$u_{tt} + u_t + u = u_{xx}$$

is what causes $u(x,t) \to 0$ as $t \to \infty$.

---

5. Given the physical interpretation of the Laplacian, what is the general nature of the solution of the following Helmholtz BVP.

$$\text{PDE: } u_{rr} + \frac{1}{r}u_r + \frac{1}{r^2}u_{\theta\theta} = -\lambda^2 u \quad 0 < r < 1$$

$$\text{BC: } u(1,\theta) = 0 \quad 0 \leq \theta < 2\pi$$

Solution: If you first look at the one-dimensional Helmoltz BVP, which is

$$\text{ODE: } \frac{d^2 u}{dx^2} = -\lambda^2 u$$

$$\text{BC: } \begin{cases} u(0) = 0 \\ u(1) = 0 \end{cases}$$

the general solution of the ODE is

$$u(x) = c_1 \sin \lambda x + c_2 \cos \lambda x$$

and plugging this solution in the BC gives

$$c_2 = 0$$
$$c_1 \sin \lambda = 0$$

We can pick $c_1 = 0$ will give us the zero solution $u(x) \equiv 0$, which we knew from the start, but we want to find nonzero solutions and so we set

$$\sin \lambda = 0$$

Lesson 32: General Nature of Boundary Value Problems

which yields values (eigenvalues) $\lambda_n = \pi, 2\pi, 3\pi, ..., n\pi, ...$ and so we find an infinite number of solutions of this one-dimensional Helmloltz problem, namely

$$u_n(x) = c\sin(n\pi x), n = 1, 2, ...$$

where c is any constant. Figure 32.1 shows the eigenfunctions $u_1(x), u_2(x), u_3(x), u_4(x)$.

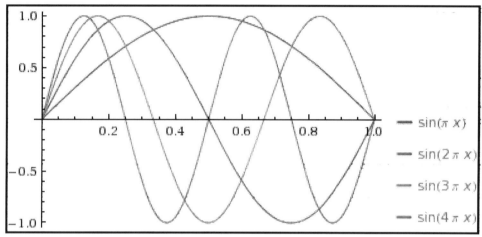

Figure 32.1 Solutions of the 1-dimensional Helmholtz eigenvalue problem

We imagine now what happens in *higher* dimensions. In two dimensions we can interpret Laplace's equation as roughly describing a stretched membrane (in the absence of gravity), where the height of every point on the membrane is the average height of values on the membrane on a small circle around the point. The Helmoltz equation is different. Whenever the height u of the membrane is positive the Laplacian is negative, and whenever $u < 0$ then $\nabla^2 u > 0$, which means when the membrane is sagging then the overall shape of the membrane is concave up, and when $u > 0$ the membrane is concave down. There will be an infinite number of these possible shapes, and they constitute the various shapes of a vibrating drumhead in the same way as the eigenfunctions in Figure 32.1 form the basic shapes of a vibrating sting.

6. What is the physical interpretation of the following mixed BVP inside the square.

PDE: $u_{xx} + u_{yy} = 0 \quad 0 < x < 1, \ 0 < y < 1$

BCs: $\begin{cases} u_y(x,0) = h[u(x,0) - 2] & 0 < x < 1 \\ u(x,1) = 1 & 0 < x < 1 \\ u_x(0,y) = 0 & 0 < y < 1 \\ u_x(1,y) = 0 & 0 < y < 1 \end{cases}$

Solution: The function $u(x,t)$ might define the temperature of a square plate, insulated on the left and right due to the BCs $u_x(0,y) = 0, u_x(1,y) = 0 \ \ 0 < y < 1$ so that heat only flows across the top and bottom sides. The temperature on the top side of the square is kept at $u(x,1) = 1$, hence heat flows in or out across that side depending on whether the temperature near that boundary is less or greater than 1, respectively (a very strong control system). On the bottom of the square, heat can enter or leave, depending on whether the temperature at that boundary is less than or greater than 2. If the temperature along the bottom edge is 1 (degree) then $u_y(x,0) < 0$ and heat will enter through the boundary (remember heat goes from hot to cold), but, if the temperature at the bottom edge (when $y = 1$) is positive, then $u_y(x,0) > 0$ and heat will leave the square. The constant h is a proportionally for this process. The solution to this problem won't be exactly 2 on the bottom edge but the larger h the closer it will be.

ΞPTOΠT

Lesson 33: Interior Dirichlet Problem for a Circle

> 1. Carry out the details or the separation of variables to the following interior Dirichlet problem.
>
> PDE: $u_{rr} + \dfrac{1}{r}u_r + \dfrac{1}{r^2}u_{\theta\theta} = 0 \quad 0 < r < 1, \quad 0 < \theta < 2\pi$
>
> BC: $u(1,\theta) = g(\theta) \quad 0 \leq \theta < 2\pi$

Solution: We look for a solution of the form $u(r,\theta) = R(r)\Theta(\theta)$, and plug this into the PDE, getting

$$R''\Theta + \frac{1}{r}R'\Theta + \frac{1}{r^2}R\Theta'' = 0$$

Now, if we divide each side of the equation by $R\Theta$, multiply by r^2, we arrive at

$$r^2 \frac{R''}{R} + r\frac{R'}{R} = -\frac{\Theta''}{\Theta}$$

Now, since the left hand side of this equation depends only on r and the right hand side only on θ, and since r and θ are independent, each side must be equal to a constant, which we call λ, and after minimal algebra, gives the two ODEs

$$r^2 R'' + rR' - \lambda R = 0$$
$$\Theta'' + \lambda\Theta = 0$$

The ODE in θ has the general solution

$$\Theta(\theta) = \begin{cases} c_1 e^{\sqrt{\lambda}\theta} + c_2 e^{-\sqrt{\lambda}\theta} & \lambda < 0 \\ a + b\theta & \lambda = 0 \\ c_1 \cos(\sqrt{\lambda}\,\theta) + c_2 \sin(\sqrt{\lambda}\,\theta) & \lambda > 0 \end{cases}$$

Since the function $\Theta(\theta)$ must be periodic in θ with period 2π, the constant λ can be any of the values (called eigenvalues) $\sqrt{\lambda_n} = n$, $n = 0, 1, 2, ...$ which give acceptable solutions

$$\Theta_n(\theta) = c_1 \cos(n\theta) + c_2 \sin(n\theta), \quad n = 0, 1, 2, ...$$

where c_1, c_2 are arbitrary constants. If we now solve the DE in R with acceptable values of $\sqrt{\lambda_n} = n$ we find

$$R(r) = \begin{cases} a + b \ln r & \lambda_n = 0 \\ c_1 r^n + c_2 r^{-n} & \sqrt{\lambda_n} = n, \ n = 1, 2, ... \end{cases}$$

The only solutions $R(r)$ that do not blow up at the origin $(r = 0)$ are the constant solution a and r^λ. We have now found an infinite number of solutions of the desired form $R(r)\Theta(\theta)$ and they are

$$u_n(r, \theta) = R_n(r)\Theta_n(\theta) = r^n \left[a_n \cos(n\theta) + b_n \sin(n\theta) \right]$$

for $n = 0, 1, 2, ...$ (note that $n = 0$ includes the constant solution. But since the PDE is linear and homogeneous we know that the sums of solutions of this type are solutions and hence sums of the form

$$u(r, \theta) = \sum_{n=0}^{\infty} r^n \left[a_n \cos(n\theta) + b_n \sin(n\theta) \right]$$

also satisfies the PDE. We now want to find the a_n, b_n so the BC is satisfied. Plugging the solution in the BC gives

$$u(1, \theta) = \sum_{n=0}^{\infty} a_n \cos(n\theta) + b_n \sin(n\theta) = g(\theta)$$

We now multiply both sides of this equation by $\cos(m\theta)$ and integrate each side of the equation with respect to θ from 0 to 2π, and using the orthogonal properties of sines and cosines

Lesson 33: Interior Dirichlet Problem for a Circle

$$\int_0^{2\pi} \sin(n\theta)\sin(m\theta)\,d\theta = \begin{cases} 0 & m \neq n \\ \pi & m = n \end{cases}$$

$$\int_0^{2\pi} \cos(n\theta)\cos(m\theta)\,d\theta = \begin{cases} 0 & m \neq n \\ \pi & m = n \end{cases}$$

$$\int_0^{2\pi} \sin(n\theta)\cos(n\theta)\,d\theta = 0 \quad n = 0,1,2,\ldots$$

we get

$$\int_0^{2\pi} \cos(m\theta) g(\theta)\,d\theta = a_m \int_0^{2\pi} \cos^2(m\theta)\,d\theta$$

or

$$a_m = \frac{\int_0^{2\pi} \cos(m\theta) g(\theta)\,d\theta}{\int_0^{2\pi} \cos^2(m\theta)\,d\theta} = \frac{1}{\pi}\int_0^{2\pi} \cos(m\theta) g(\theta)\,d\theta,\ n = 1,2,\ldots$$

We find the coefficients b_n by multiplying each side of the above equation by $\sin(n\theta)$ and integrating, arriving at

$$b_n = \frac{\int_0^{2\pi} \sin(n\theta) g(\theta)\,d\theta}{\int_0^{2\pi} \sin^2(n\theta)\,d\theta} = \frac{1}{\pi}\int_0^{2\pi} \sin(n\theta) g(\theta)\,d\theta,\ n = 0,1,2,\ldots$$

---

2. What is the solution of the interior Dirichlet problem

PDE: $u_{rr} + \dfrac{1}{r} u_r + \dfrac{1}{r^2} u_{\theta\theta} = 0 \quad 0 < r < 1,\ 0 < \theta < 2\pi$

BC: $u(1,\theta) = g(\theta) \quad 0 \leq \theta < 2\pi$

for the following BCs. What do the solutions look like?

a) $u(1,\theta) = 1 + \sin\theta + \dfrac{1}{2}\cos\theta$

b) $u(1,\theta) = 2$

c) $u(1,\theta) = \sin\theta$

d) $u(1,\theta) = \sin(3\theta)$

Solution:

a) From a physical point of view a stretched rubber sheet is fixed on the boundary at

$$u(1,\theta) = 1 + \sin\theta + \frac{1}{2}\cos\theta$$

We can think of decomposing this problem into 3 parts, each one with one of the terms on the boundary. A value of $u(1,\theta) = 1$ on the boundary gives the solution of $u(x,\theta) = 1$ inside the circle (clearly any constant satisfies Laplace's equation since all terms are derivatives).

The BCs of $\sin\theta$ and $\cos\theta$ give rise to the stretched rubber sheet being linear in r. The conclusion is that the solution to the given BC is

$$u(r,\theta) = 1 + r\sin\theta + \frac{1}{2}r\cos\theta$$

You can verify that this function satisfies both the PDE and BC.

b) $u(r,\theta) = 2$

c) $u(r,\theta) = r\sin\theta$

d) This solution is more complicated. If you envision a stretched rubber sheet being fixed at $u(1,\theta) = \sin(3\theta)$ on the boundary, the r dependence is more complicated. It is not linear in r as it was for $\sin\theta$ and $\cos\theta$. You might be temped to guess that if $u(r,\theta) = r^3 \sin(3\theta)$ is the solution of Laplace's equation with BC $u(1,\theta) = \sin(3\theta)$, and that

$$u(1,\theta) = \sin(\theta) \Rightarrow u(r,\theta) = r\sin\theta$$
$$u(1,\theta) = \sin(2\theta) \Rightarrow u(r,\theta) = r^2 \sin\theta$$
$$u(1,\theta) = \sin(3\theta) \Rightarrow u(r,\theta) = r^3 \sin\theta$$
$$\ldots \quad \ldots \quad \ldots$$

If you guessed this, you would be right. You can also verify that $u(r,\theta) = r^3 \sin(3\theta)$ satisfies Laplace's equation.

Lesson 33: Interior Dirichlet Problem for a Circle

3. What is the solution of the following interior Dirichlet problem when the radius of the circle is $R = 2$.

$$\text{PDE:} \quad u_{rr} + \frac{1}{r}u_r + \frac{1}{r^2}u_{\theta\theta} = 0 \quad 0 < r < 2, \; 0 < \theta < 2\pi$$

$$\text{BC:} \quad u(2,\theta) = g(\theta) \quad 0 \leq \theta < 2\pi$$

Solution: For an integral type solution, we simply plug in $R = 2$ in the general Poisson integral formula getting

$$u(r,\theta) = \frac{1}{2\pi}\int_0^{2\pi}\left[\frac{R^2 - r^2}{R^2 + 2rR\cos(\xi - \omega) + r^2}\right]g(\omega)d\omega$$

$$= \frac{1}{2\pi}\int_0^{2\pi}\left[\frac{4 - r^2}{4 + 4r\cos(\xi - \omega) + r^2}\right]g(\omega)d\omega$$

There is also a Fourier series type solution. We have solved this Dirichlet problem for $0 < r < 1$, but we can modify the solution for $0 < r < 2$ by making a change of variable $\xi = r/2$, where now

$$0 < r < 2 \Rightarrow 0 < \xi < 1$$

We can now use the solution with radius 1 in terms of $u(\xi,\theta)$ and then plug in $\xi = r/2$ in this solution. The solution for

$$\text{PDE:} \quad u_{\xi\xi} + \frac{1}{\xi}u_\xi + \frac{1}{\xi^2}u_{\theta\theta} = 0 \quad 0 < \xi < 1, \; 0 < \theta < 2\pi$$

$$\text{BC:} \quad u(1,\theta) = g(\theta) \quad 0 \leq \theta < 2\pi$$

is
$$u(\xi,\theta) = \sum_{n=1}^{\infty} a_n \xi^n \sin(n\pi\theta)$$

where

Lesson 33: Interior Dirichlet Problem for a Circle

$$a_n = \frac{\int_0^1 \sin(n\pi\theta) g(\theta) d\theta}{\int_0^1 \sin^2(n\pi\theta) d\theta} = 2\int_0^1 \sin(n\pi\theta) g(\theta) d\theta$$

Hence, the solution in terms of r, θ is

$$u(r,\theta) = \sum_{n=1}^{\infty} a_n \left(\frac{r}{2}\right)^n \sin(n\pi\theta), \quad 0 \leq r \leq 2, \quad 0 < \theta \leq 2\pi$$

where the coefficients a_n are the same values as given above.

---

4. What is the solution of the following interior Dirichlet problem with radius $R = 2$ if the BC are $u(2, \theta) = \sin(2\theta)$.

PDE: $u_{rr} + \frac{1}{r} u_r + \frac{1}{r^2} u_{\theta\theta} = 0 \quad 0 < r < 2, \quad 0 < \theta < 2\pi$

BC: $u(2, \theta) = \sin(2\theta) \quad 0 \leq \theta < 2\pi$

Solution: We essentially solved this problem in Problem 3, but we will do it again here in slightly more detail. We can transform this problem to a problem on a circle of radius 1, solve the problem with radius 1, and then transform back to this problem. We do this by introducing a new variable

$$r = 2\xi$$

Then since

$$u_\xi = u_r r_\xi = 2u_r \Rightarrow u_r = \frac{1}{2} u_\xi$$

$$u_{\xi\xi} = \frac{\partial}{\partial \xi}[2u_r] = 2u_{rr} r_\xi = 4u_{rr} \Rightarrow u_{rr} = \frac{1}{4} u_{\xi\xi}$$

the new transformed problem becomes

PDE: $\frac{1}{4} u_{\xi\xi} + \frac{1}{4\xi} u_\xi + \frac{1}{4\xi^2} u_{\theta\theta} = 0 \quad 0 < \xi < 1, \quad 0 < \theta < 2\pi$

BC: $u(1, \theta) = \sin(2\theta) \quad 0 \leq \theta < 2\pi$

Lesson 33: Interior Dirichlet Problem for a Circle

or

$$\text{PDE: } u_{\xi\xi} + \frac{1}{\xi}u_\xi + \frac{1}{\xi^2}u_{\theta\theta} = 0 \quad 0<\xi<1, \quad 0<\theta<2\pi$$

$$\text{BC: } u(1,\theta) = \sin(2\theta) \quad 0 \leq \theta < 2\pi$$

The transformed problem has the solution
$$u(\xi,\theta) = \xi^2 \sin(2\theta)$$
and so
$$u(r,\theta) = \frac{r^2}{4}\sin(2\theta)$$

You can verify this function satisfies the Dirichlet problem on a circle of radius $R=4$. In a nutshell we went from a circle of radius 1 to radius 2 by replacing r in the solution by $r/2$.

---

5. Solve the following interior Dirichlet problem

$$\text{PDE: } u_{rr} + \frac{1}{r}u_r + \frac{1}{r^2}u_{\theta\theta} = 0 \quad 0<r<2, \quad 0<\theta<2\pi$$

$$\text{BC: } u(2,\theta) = \begin{cases} \sin\theta & 0 \leq \theta < \pi \\ 0 & \pi \leq \theta < 2\pi \end{cases}$$

Solution: We have seen that in polar coordinates, Laplace's equation has solutions of the form

$$u(r,\theta) = \sum_{n=1}^{\infty} r^n \left[a_n \cos(n\theta) + b_n \sin(n\theta) \right]$$

To find the coefficients a_n, b_n, we plug this solution in the BC, getting

$$u(2,\theta) = \sum_{n=1}^{\infty} 2^n \left[a_n \cos(n\theta) + b_n \sin(n\theta) \right] = \begin{cases} \sin\theta & 0 \leq \theta < \pi \\ 0 & \pi \leq \theta < 2\pi \end{cases}$$

To find the a_n we multiply both sides of the equation by $\cos(m\theta)$ and integrate each side of the equation from 0 to 2π, getting

Lesson 33: Interior Dirichlet Problem for a Circle

$$2^m a_m \int_0^{2\pi} \cos^2(m\theta)\, d\theta = \int_0^{\pi} \sin\theta \cos(m\theta)\, d\theta$$

or

$$a_m = \frac{\int_0^{\pi} \sin\theta \cos(m\theta)\, d\theta}{2^m \int_0^{2\pi} \cos^2(m\theta)\, d\theta} = \begin{cases} 0 & m = 1, 3, 5, \ldots \\ \left(\dfrac{1}{2^m}\right)\left(\dfrac{1}{1-m^2}\right), & m = 0, 2, 4, \ldots \end{cases}$$

(these integrals were carried out on Mathematica, the reader should have access to a variety of computer algebra systems). We find the coefficients b_m in a similar way.

---

6. What does the Poisson kernel

$$\text{Poisson kernel} = \frac{R^2 - r^2}{R^2 - 2rR\cos(\theta - \alpha) + r^2}$$

look like as a function of α, $0 \leq \alpha < 2\pi$ for

$$r = \frac{3R}{4},\ \theta = \frac{\pi}{2}$$

Solution: Picking $R = 1$ we graph the following Poisson kernel using Mathematica.

$$\text{Poisson kernel} = \frac{R^2 - r^2}{R^2 - 2rR\cos(\theta - \alpha) + r^2}$$

$$= \frac{1 - (9/16)}{1 - 1.5\cos\left(\dfrac{\pi}{2} - \alpha\right) + \left(\dfrac{9}{16}\right)},\quad 0 \leq \alpha \leq 2\pi$$

Lesson 33: Interior Dirichlet Problem for a Circle

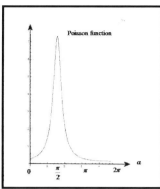

Figure 33.1 Poisson's function

7. If $V(r,\theta)$ is *any* solution of the PDE (not including the BC)

$$\text{PDE:} \quad V_{rr} + \frac{1}{r}V_r + \frac{1}{r^2}V_{\theta\theta} = f(r,\theta) \quad 0 < r < 1, \quad 0 < \theta < 2\pi$$

and if $W(r,\theta)$ is a solution of

$$\text{PDE:} \quad W_{rr} + \frac{1}{r}W_r + \frac{1}{r^2}W_{\theta\theta} = 0 \quad 0 < r < 1, \quad 0 < \theta < 2\pi$$

$$\text{BC:} \quad W(1,\theta) = V(1,\theta) \quad 0 \leq \theta < 2\pi$$

then the solution $u(r,\theta)$ of

$$\text{PDE:} \quad u_{rr} + \frac{1}{r}u_r + \frac{1}{r^2}u_{\theta\theta} = f(r,\theta) \quad 0 < r < 1, \quad 0 < \theta < 2\pi$$

$$\text{BC:} \quad u(1,\theta) = 0 \quad 0 \leq \theta < 2\pi$$

is

$$u(r,\theta) = V(r,\theta) - W(r,\theta)$$

Note: The note in the book was stated more generally, but we state a restrictive version here so the proof is more concrete.

Solution: Plugging

$$u(r,\theta) = V(r,\theta) - W(r,\theta)$$

into

PDE: $u_{rr} + \dfrac{1}{r}u_r + \dfrac{1}{r^2}u_{\theta\theta} = f(r,\theta)$ $0<r<1$, $0<\theta<2\pi$

BC: $u(1,\theta) = 0$ $0 \le \theta < 2\pi$

gives

PDE: $\dfrac{\partial^2(V-W)}{\partial r^2} + \dfrac{1}{r}\dfrac{\partial(V-W)}{\partial r} + \dfrac{1}{r^2}\dfrac{\partial^2(V-W)}{\partial \theta^2}$

$= \left[\dfrac{\partial^2 V}{\partial r^2} + \dfrac{1}{r}\dfrac{\partial V}{\partial r} + \dfrac{1}{r^2}\dfrac{\partial^2 V}{\partial \theta^2}\right] - \left[\dfrac{\partial^2 W}{\partial r^2} + \dfrac{1}{r}\dfrac{\partial W}{\partial r} + \dfrac{1}{r^2}\dfrac{\partial^2 W}{\partial \theta^2}\right]$

$= f(r,\theta) - 0 = f(r,\theta)$

BC: $V(1,\theta) - W(1,\theta) = V(1,\theta) - V(1,\theta) = 0$

---

8. If $V(r,\theta)$ is *any* function that satisfies the BC $V(1,\theta) = f(\theta)$ for a given $f(\theta)$, and if the function $W(r,\theta)$ satisfies the BVP

PDE: $W_{rr} + \dfrac{1}{r}W_r + \dfrac{1}{r^2}W_{\theta\theta} = V_{rr} + \dfrac{1}{r}V_r + \dfrac{1}{r^2}V_{\theta\theta}$ $0<r<1$, $0<\theta<2\pi$

BC: $W(1,\theta) = 0$ $0 \le \theta < 2\pi$

then

$$u(r,\theta) = V(r,\theta) - W(r,\theta)$$

satisfies

PDE: $u_{rr} + \dfrac{1}{r}u_r + \dfrac{1}{r^2}u_{\theta\theta} = 0$ $0<r<1$, $0<\theta<2\pi$

BC: $u(1,\theta) = f(\theta)$ $0 \le \theta < 2\pi$

Note: The note in the book was stated more generally, but we state a more restrictive version here so the proof if more concrete.

Solution: We are given that $W(r,\theta)$ satisfies the BVP

Lesson 33: Interior Dirichlet Problem for a Circle

PDE: $W_{rr} + \dfrac{1}{r}W_r + \dfrac{1}{r^2}W_{\theta\theta} = V_{rr} + \dfrac{1}{r}V_r + \dfrac{1}{r^2}V_{\theta\theta}$ $0 < r < 1$, $0 < \theta < 2\pi$

BC: $W(1,\theta) = 0$ $0 \le \theta < 2\pi$

where $V(r,\theta)$ is an arbitrary (differentiable) function that satisfies $V(1,\theta) = f(\theta)$. From these equations, we have

PDE: $\dfrac{\partial^2(V-W)}{\partial r^2} + \dfrac{1}{r}\dfrac{\partial(V-W)}{\partial r} + \dfrac{1}{r^2}\dfrac{\partial^2(V-W)}{\partial \theta^2} = 0$

BC: $V(1,\theta) - W(1,\theta) = f(\theta) - 0 = f(\theta)$

which proves the desired result.

ΞΤΨΘΠΥ

Lesson 34: The Dirichlet Problem in an Annulus

1. Solve the Dirichlet problem in an annulus

PDE: $u_{rr} + \dfrac{1}{r}u_r + \dfrac{1}{r^2}u_{\theta\theta} = 0 \quad 1 < r < 2, \quad 0 < \theta < 2\pi$

BCs: $\begin{cases} u(1,\theta) = \cos\theta \\ u(2,r) = \sin\theta \end{cases} \quad 0 \le \theta < 2\pi$

Solution: Due to the BCs we know the general form of the Dirichlet problem inside an annulus has the general form

$$u(r,\theta) = \sum_{n=1}^{\infty}\left[\left(a_n r^n + b_n r^{-n}\right)\cos(n\theta) + \left(c_n r^n + d_n r^{-n}\right)\sin(n\theta)\right]$$

it is clear the only nonzero coefficients are a_1, b_1, c_1, and d_1. To find these coefficients, we choose them so u satisfies the BCs, which result in the following four equations, the first 2 in a_1, b_1, the second 2 in c_1, d_1:

$$a_1 + b_1 = \frac{1}{\pi}\int_0^{2\pi} \cos^2(\theta)\,d\theta = 1$$

$$4a_1 + \frac{1}{4}b_1 = \frac{1}{\pi}\int_0^{2\pi} \cos(\theta)\sin(\theta)\,d\theta = 0$$

$$c_1 + d_1 = \frac{1}{\pi}\int_0^{2\pi} \sin(\theta)\cos(\theta)\,d\theta = 0$$

$$4c_1 + \frac{1}{4}d_1 = \frac{1}{\pi}\int_0^{2\pi} \sin^2(\theta)\,d\theta = 1$$

Solving these equations results in

$$a_1 = -\frac{1}{3},\ b_1 = \frac{4}{3},\ c_1 = \frac{2}{3},\ d_1 = -\frac{2}{3}$$

Lesson 34: The Dirichlet Problem in an Annulus

Hence, we have the solution

$$u(r,\theta) = \left[-\frac{r}{3} + \frac{4}{3r}\right]\cos\theta + \left[\frac{2r}{3} - \frac{2}{3r}\right]\sin\theta$$

---

2. What is the solution to the exterior Dirichlet problem

PDE: $u_{rr} + \frac{1}{r}u_r + \frac{1}{r^2}u_{\theta\theta} = 0 \quad 1 < r < \infty, \quad 0 < \theta < 2\pi$

BC: $u(1,\theta) = g(\theta) \quad 0 \leq \theta < 2\pi$

for the following BCs:

a) $g(\theta) = 1$

b) $g(\theta) = 1 + \cos(3\theta)$

c) $g(\theta) = \sin\theta + \cos(3\theta)$

d) $g(\theta) = \begin{cases} 1 & 0 \leq \theta < \pi \\ 0 & \pi \leq \theta < 2\pi \end{cases}$

Solution: The general solution of Laplace's equation in an annulus is

$$u(r,\theta) = a_0 + b_0 \ln r + \sum_{n=1}^{\infty}\left[\left(a_n r^n + b_n r^{-n}\right)\cos(n\theta) + \left(c_n r^n + d_n r^{-n}\right)\sin(n\theta)\right]$$

hence we simply assign the coefficients a_n, b_n, $n = 0,1,2,\ldots$ to match the BCs. Doing this, we find

a) $u(r,\theta) = 1$

b) $u(r,\theta) = 1 + \frac{1}{r^3}\cos(3\theta)$

c) $u(r,\theta) = \frac{1}{r}\sin\theta + \frac{1}{r^3}\cos(3\theta)$

d) For the exterior Dirichlet problem, the form of the solution is

$$u(r,\theta) = \sum_{n=0}^{\infty} r^{-n}\left[a_n \cos(n\theta) + b_n \sin(n\theta)\right]$$

$$= a_0 + \frac{1}{r}\left[a_1 \cos(\theta) + b_1 \sin(\theta)\right] + \frac{1}{r^2}\left[a_2 \cos(2\theta) + b_2 \sin(2\theta)\right] + \cdots$$

Plugging this series in the BC gives

$$u(1,\theta) = \sum_{n=0}^{\infty}\left[a_n \cos(n\theta) + b_n \sin(n\theta)\right] = \begin{cases} 1 & 0 \le \theta < \pi \\ 0 & \pi \le \theta < 2\pi \end{cases}$$

To find the coefficients a_n we multiply both sides of this equation by $\cos(m\theta)$ and integrate each side of the equation with respect to θ from 0 to 2π, getting

$$a_m \int_0^{2\pi} \cos^2(m\theta)\,d\theta = \int_0^{\pi} \cos(m\theta)\,d\theta$$

which gives

$$a_m = \frac{\int_0^{\pi} \cos(m\theta)\,d\theta}{\int_0^{2\pi} \cos^2(m\theta)\,d\theta} = 0$$

Computing b_m by the same strategy, we find

$$a_m = \frac{\int_0^{\pi} \sin(m\theta)\,d\theta}{\int_0^{2\pi} \sin^2(m\theta)\,d\theta} = \begin{cases} 0 & m = 0,2,4,\ldots \\ \dfrac{2}{m} & m = 1,3,5,\ldots \end{cases}$$

Hence, the solution is

$$u(r,\theta) = 2 \sum_{n=1,3,\ldots}^{\infty} \frac{1}{nr^n} \sin(n\theta)$$

$$= 2\left[\frac{1}{r}\sin(\theta) + \frac{1}{3r^3}\sin(3\theta) + \frac{1}{5r^5}\sin(5\theta) + \cdots\right]$$

Lesson 34: The Dirichlet Problem in an Annulus

for $1 \leq r < \infty$, $0 \leq \theta < 2\pi$.

---

3. The solution of the exterior Neumann problem

$$\text{PDE: } u_{rr} + \frac{1}{r}u_r + \frac{1}{r^2}u_{\theta\theta} = 0 \quad 1 < r < \infty, \quad 0 < \theta < 2\pi$$

$$\text{BC: } \frac{\partial u}{\partial r}(1,\theta) = g(\theta) \quad 0 \leq \theta < 2\pi$$

has the same form as the exterior Dirichlet problem, that is

$$u(r,\theta) = \sum_{n=1}^{\infty} r^{-n}\left[a_n \cos(n\theta) + b_n \sin(n\theta)\right]$$

where the constants a_n, b_n are chosen so that $u(r,\theta)$ satisfies the BC. Find the solution of the above Neumann problem with BC

$$\frac{\partial u}{\partial r}(1,\theta) = \sin\theta$$

Solution: Computing the derivative (with respect to r) of the general form, we have

$$u_r(r,\theta) = \sum_{n=1}^{\infty} -n r^{-(n+1)}\left[a_n \cos(n\theta) + b_n \sin(n\theta)\right]$$

and plugging this into the BC we get

$$u_r(1,\theta) = -\sum_{n=1}^{\infty} n\left[a_n \cos(n\theta) + b_n \sin(n\theta)\right] = g(\theta)$$

We find the coefficients a_n, b_n in the usual manner. To find a_n we multiply each side of this equation by $\cos(m\theta)$ and integrate each side of the equation from 0 to 2π, getting

$$-m a_m = \int_0^{2\pi} g(\theta)\cos(m\theta)d\theta = \int_0^{2\pi} \sin(m\theta)\cos(m\theta)d\theta = 0$$

or $a_m = 0$. To find b_n we multiply by $\sin(m\theta)$ and integrate, getting

$$-mb_m \int_0^{2\pi} \sin^2(m\theta) = \int_0^{2\pi} \sin(m\theta) g(\theta) d\theta$$

or

$$b_m = -\frac{\int_0^{2\pi} \sin(m\theta) g(\theta) d\theta}{m \int_0^{2\pi} \sin^2(m\theta)}$$

$$= -\frac{\int_0^{2\pi} \sin(m\theta) \sin(\theta) d\theta}{m \int_0^{2\pi} \sin^2(m\theta)}$$

$$= \begin{cases} 0 & m \neq 1 \\ -1 & m = 1 \end{cases}$$

Hence, the solution

$$u(r,\theta) = -\frac{1}{r}\sin(\theta)$$

Checking the solution we find it satisfies the BC

$$u_r(r,\theta) = \frac{1}{r^2}\sin(\theta) \Rightarrow u_r(1,\theta) = \sin(\theta)$$

---

4. Substitute the general solution

$$u(r,\theta) = a_0 + b_0 \ln r + \sum_{n=1}^{\infty} \left[\left(a_n r^n + b_n r^{-n}\right)\cos(n\theta) + \left(c_n r^n + d_n r^{-n}\right)\sin(n\theta) \right]$$

in the BC

$$u(R_1,\theta) = g_1(\theta)$$
$$u(R_2,\theta) = g_2(\theta)$$

and integrating with respect to θ gives the following equations from which we can find a_n, b_n, $n = 0, 1, 2, \ldots$,

Lesson 34: The Dirichlet Problem in an Annulus

$$\begin{cases} a_0 + b_0 \ln R_1 = \dfrac{1}{2\pi} \int_0^{2\pi} g_1(\theta) d\theta \\ a_0 + b_0 \ln R_2 = \dfrac{1}{2\pi} \int_0^{2\pi} g_2(\theta) d\theta \end{cases}$$

$$\begin{cases} a_n R_1^n + b_n R_1^{-n} = \dfrac{1}{\pi} \int_0^{2\pi} g_1(\theta) \cos(n\theta) d\theta \\ a_n R_2^n + b_n R_2^{-n} = \dfrac{1}{\pi} \int_0^{2\pi} g_2(\theta) \cos(n\theta) d\theta \end{cases}$$

$$\begin{cases} c_n R_1^n + d_n R_1^{-n} = \dfrac{1}{\pi} \int_0^{2\pi} g_1(\theta) \sin(n\theta) d\theta \\ c_n R_2^n + d_n R_2^{-n} = \dfrac{1}{\pi} \int_0^{2\pi} g_2(\theta) \sin(n\theta) d\theta \end{cases}$$

Solution: To find a_0, b_0 we evaluate the BCs $u(r,\theta) = g(\theta)$ at $r = R_1, R_2$ getting

$$u(R_1, \theta) = a_0 + b_0 \ln R_1 + \sum_{n=1}^{\infty} \left[\left(a_n R_1^n + b_n R_1^{-n} \right) \cos(n\theta) + \left(c_n R_1^n + d_n R_1^{-n} \right) \sin(n\theta) \right] = g_1(\theta)$$

$$u(R_2, \theta) = a_0 + b_0 \ln R_2 + \sum_{n=1}^{\infty} \left[\left(a_n R_2^n + b_n R_2^{-n} \right) \cos(n\theta) + \left(c_n R_2^n + d_n R_2^{-n} \right) \sin(n\theta) \right] = g_2(\theta)$$

then integrating each equation with respect to θ from 0 to 2π, and using the fact that

$$\int_0^{2\pi} \sin(n\theta) d\theta = 0, \quad n = 1, 2, \ldots$$

$$\int_0^{2\pi} \cos(n\theta) d\theta = 0, \quad n = 1, 2, \ldots$$

we get

$$a_0 + b_0 \ln R_1 = \frac{1}{2\pi}\int_0^{2\pi} g_1(\theta)d\theta$$

$$a_0 + b_0 \ln R_2 = \frac{1}{2\pi}\int_0^{2\pi} g_2(\theta)d\theta$$

To find the coefficients a_n, b_n we evaluate the BCs $u(r,\theta) = g(\theta)$ at $r = R_1, R_2$, getting the equations

$$u(R_1,\theta) = a_0 + b_0 \ln R_1 + \sum_{n=1}^{\infty}\left[\left(a_n R_1^n + b_n R_1^{-n}\right)\cos(n\theta) + \left(c_n R_1^n + d_n R_1^{-n}\right)\sin(n\theta)\right] = g_1(\theta)$$

$$u(R_2,\theta) = a_0 + b_0 \ln R_2 + \sum_{n=1}^{\infty}\left[\left(a_n R_2^n + b_n R_2^{-n}\right)\cos(n\theta) + \left(c_n R_2^n + d_n R_2^{-n}\right)\sin(n\theta)\right] = g_2(\theta)$$

Now multiply each equation by $\cos(m\theta)$ and integrate each equation with respect to θ from 0 to 2π, and using the orthogonality properties

$$\int_0^{2\pi}\sin(m\theta)\cos(n\theta)d\theta = 0, \quad n = 1,2,\ldots$$

$$\int_0^{2\pi}\cos(m\theta)\cos(n\theta)d\theta = \begin{cases} 0 & m \neq n \\ \pi & m = n \end{cases}$$

we (simple exercise) arrive at the equations

$$a_n R_1^n + b_n R_1^{-n} = \frac{1}{\pi}\int_0^{2\pi} g_1(\theta)\cos(n\theta)d\theta$$

$$a_n R_2^n + b_n R_2^{-n} = \frac{1}{\pi}\int_0^{2\pi} g_2(\theta)\cos(n\theta)d\theta$$

Finally, to find the coefficients c_n, d_n we do the same thing as we did when we found a_n, b_n except now we multiply the equations by $\sin(m\theta)$. We leave this simple exercise for the reader.

ΞΡΠΠΘΖ

Lesson 35: Laplace's Equation in Spherical Coordinates (Spherical Harmonics)

1. Find two linear independent solutions of Euler's

$$r^2 \frac{d^2R}{dr^2} + 2r\frac{dR}{dr} - n(n+1)R = 0$$

of the form $R(r) = r^\alpha$.

Solution: Plugging

$$R(r) = r^\alpha$$

in Euler's equation gives

$$r^2 \left[\alpha(\alpha-1)r^{(\alpha-2)}\right] + 2r\left[\alpha r^{(\alpha-1)}\right] - n(n+1)r^\alpha = 0$$

or

$$r^2 \left[\alpha(\alpha-1)r^{(\alpha-2)}\right] + 2r\left[\alpha r^{(\alpha-1)}\right] - n(n+1)r^\alpha = 0$$

$$\therefore \ \alpha(\alpha-1)r^\alpha + 2\alpha r^\alpha - n(n+1)r^\alpha = 0$$

$$\therefore \ r^\alpha \left[\alpha(\alpha-1) + 2\alpha - n(n+1)\right] = 0$$

$$\therefore \ \alpha^2 + \alpha - n(n+1) = 0$$

This last equation is a quadratic equation in α, whose solutions can easily be found to be $\alpha = n, -(n+1)$. Hence, we have found two linearly independent solutions of Euler's equation

$$R_1(r) = r^n, \ R_2(r) = \frac{1}{r^{n+1}}, \ n = 0, 1, 2, \ldots$$

which yields the general solution

$$R(r) = c_1 R_1(r) + c_2 R_2(r) = c_1 r^n + c_2 \left(\frac{1}{r^{n+1}}\right), \ n = 0, 1, 2, \ldots$$

> **2.** Make the change of variables $x = \cos\phi$ to change the Legendre's equation in independent variable ϕ
>
> $$\frac{d}{d\phi}\left[\sin\phi \frac{d\Phi}{d\phi}\right] + n(n+1)\sin\phi\, \Phi = 0 \quad 0 < \phi < \pi$$
>
> to the following one in independent variable x
>
> $$(1-x^2)\frac{d^2\Phi}{dx^2} - 2x\frac{d\Phi}{dx} + n(n+1)\Phi = 0, \quad -1 < x < 1$$

Solution: First realize that Legendre's equation

$$\frac{d}{d\phi}\left[\sin\phi \frac{d\Phi}{d\phi}\right] + n(n+1)\sin\phi\, \Phi = 0 \quad 0 < \phi < \pi$$

can be rewritten in the alternate form

$$\sin\phi \frac{d^2\Phi}{d\phi^2} + \cos\phi \frac{d\Phi}{d\phi} + n(n+1)\sin\phi\, \Phi = 0$$

by simple differentiation. Now, by making the change of variables $x = \cos\phi$ we see

Lesson 35: Laplace's Equation in Spherical Coordinates

$$\frac{d\Phi}{d\phi} = \frac{d\Phi}{dx}\frac{dx}{d\phi} = -\sin\phi\frac{d\Phi}{dx}$$

$$\frac{d^2\Phi}{d\phi^2} = \frac{d}{d\phi}\left(\frac{d\Phi}{d\phi}\right)$$

$$= \frac{d}{d\phi}\left(-\sin\phi\frac{d\Phi}{dx}\right)$$

$$= (-\cos\phi)\frac{d\Phi}{dx} - \sin\phi\frac{d}{d\phi}\left(\frac{d\Phi}{dx}\right)$$

$$= (-\cos\phi)\frac{d\Phi}{dx} - \sin\phi\left[\frac{d}{dx}\left(\frac{d\Phi}{dx}\right)\left(\frac{dx}{d\phi}\right)\right]$$

$$= (-\cos\phi)\frac{d\Phi}{dx} - \sin\phi\left[\frac{d^2\Phi}{dx^2}(-\sin\phi)\right]$$

$$= \sin^2\phi\frac{d^2\Phi}{dx^2} - \cos\phi\frac{d\Phi}{dx}$$

Hence, Legendre's equation

$$\sin\phi\frac{d^2\Phi}{d\phi^2} + \cos\phi\frac{d\Phi}{d\phi} + n(n+1)\sin\phi\,\Phi = 0$$

is transformed to

$$\sin\phi \frac{d^2\Phi}{d\phi^2} + \cos\phi \frac{d\Phi}{d\phi} + n(n+1)\sin\phi\,\Phi$$

$$= \sin\phi\left[\sin^2\phi \frac{d^2\Phi}{dx^2} - \cos\phi \frac{d\Phi}{dx}\right] + \cos\phi\left[-\sin\phi \frac{d\Phi}{dx}\right] + n(n+1)\sin\phi\,\Phi$$

$$= \sin^3\phi \frac{d^2\Phi}{dx^2} - 2\sin\phi\cos\phi \frac{d\Phi}{dx} + n(n+1)\sqrt{1-x^2}\,\Phi$$

$$= \sqrt{1-x^2}\,(1-x^2)\frac{d^2\Phi}{d\phi^2} - 2x\sqrt{1-x^2}\frac{d\Phi}{dx} + n(n+1)\sqrt{1-x^2}\,\Phi$$

$$= \sqrt{1-x^2}\left[(1-x^2)\frac{d^2\Phi}{d\phi^2} - 2x\frac{d\Phi}{dx} + n(n+1)\Phi\right]$$

Hence Lejendre's equation

$$\sin\phi \frac{d^2\Phi}{d\phi^2} + \cos\phi \frac{d\Phi}{d\phi} + n(n+1)\sin\phi\,\Phi = 0$$

in ϕ holds, if and only if, Legendre's equation

$$(1-x^2)\frac{d^2\Phi}{d\phi^2} - 2x\frac{d\Phi}{dx} + n(n+1)\Phi = 0$$

in x holds.

---

3. Verify Rodrigues' formula

$$P_n(x) = \frac{1}{2^n n!} \frac{d^n}{dx^n}\left[(x^2-1)^n\right] \quad n=0,1,2,\ldots$$

for the Legendre polynomials $P_0(x), P_1(x), P_2(x),$ and $P_3(x)$.

Solution: By direct computation

Lesson 35: Laplace's Equation in Spherical Coordinates

$$P_0(x) = \frac{1}{2^0 0!} \frac{d^0}{dx^0}\left[(x^2-1)^0\right] = 1, \quad -1 \le x \le 1$$

$$P_1(x) = \frac{1}{2^1 1!} \frac{d}{dx}\left[(x^2-1)^1\right] = \frac{1}{2}(2x) = x, \quad -1 \le x \le 1$$

$$P_2(x) = \frac{1}{2^2 2!} \frac{d^2}{dx^2}\left[(x^2-1)^2\right] = \frac{1}{2}(3x^2-1), \quad -1 \le x \le 1$$

$$P_3(x) = \frac{1}{2^3 3!} \frac{d^3}{dx^3}\left[(x^2-1)^3\right] = \frac{1}{2}(5x^3-3x), \quad -1 \le x \le 1$$

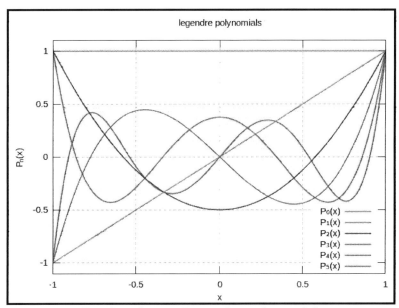

Graphs of some Legendre polynomials

---

4. Solve the interior Dirichlet problem

PDE: $(r^2 u_r)_r + \frac{1}{\sin\phi}\left[\sin\phi\, u_\phi\right]_\phi + \frac{1}{\sin^2\phi} u_{\theta\theta} = 0 \quad 0 < r < 1$

BC: $u(1,\phi) = \cos(3\phi) \quad 0 \le \phi < \pi$

Hint: Use a trig formula to write $\cos(3\phi)$ in terms of $\cos\phi, \cos^2\phi, \cos^3\phi$ and then use them in the expansion

$$\cos(3\phi) = a_0 P_0(\cos\phi) + a_1 P_1(\cos\phi) + \cdots$$

Solution: Since the BC only depends on the co-latitude ϕ, the solution has the general form

$$u(r,\phi) = \sum_{n=0}^{\infty} a_n r^n P_n(\cos\phi)$$

so the idea is to expand the BC

$$u(1,\phi) = \sum_{n=0}^{\infty} a_n P_n(\cos\phi)$$

as a series of Legendre polynomials, and then slip the factor r^n in the *nth* term of this series. For our BC we use the trig identity

$$\cos(3\phi) = 4\cos^3\phi - 3\cos\phi$$

$$= \left(\frac{8}{5}\right)\left(\frac{1}{2}\right)\left[5\cos^3\phi - 3\cos\phi\right] - \frac{3}{5}\cos\phi$$

$$= \frac{8}{5}P_3(\cos\phi) - \frac{3}{5}P_1(\cos\phi)$$

and so the solution is

$$u(r,\phi) = \frac{8}{5}r^3 P_3(\cos\phi) - \frac{3}{5}r P_1(\cos\phi)$$

$$= \frac{4}{5}r^3\left[5\cos^3\phi - 3\cos\phi\right] - \frac{3}{5}r\cos\phi$$

$$= 4r^3\cos^3\phi - \frac{1}{5}(12r^3 - 3r)\cos\phi$$

Solve the Dirichlet problem

PDE: $(r^2 u_r)_r + \dfrac{1}{\sin\phi}\left[\sin\phi\, u_\phi\right]_\phi + \dfrac{1}{\sin^2\phi} u_{\theta\theta} = 0 \quad 0 < r < 1$

BC: $u(1,\phi) = \begin{cases} 1 & 0 \le \phi < \pi/2 \\ -1 & \pi/2 \le \phi < \pi \end{cases}$

Solution: We saw in the book the solution is of Laplace's equation in spherical coordinates that does not depend on the longitude θ, but only on r and the co-latitude φ is

Lesson 35: Laplace's Equation in Spherical Coordinates

$$u(r,\phi) = \sum_{n=0}^{\infty} a_n r^n P_n(\cos\phi)$$

where the coefficients a_n are chosen so u satisfies the BC. In other words, satisfies the equation

$$u(1,\phi) = \sum_{n=1}^{\infty} a_n P_n(\cos\phi) = \begin{cases} 1 & 0 \le \phi < \pi/2 \\ -1 & \pi/2 \le \phi < \pi \end{cases}$$

To find the coefficients a_n we multiply both sides of this equation by $P_m(\cos\phi)\sin\phi$ and integrate with respect to ϕ from 0 to π, getting

$$\sum_{n=0}^{\infty} a_n \int_0^{\pi} P_n(\cos\phi) P_m(\cos\phi) \sin\phi \, d\phi = \int_0^{\pi} u(1,\phi) P_m(\cos\phi) \sin\phi \, d\phi$$

and using the orthogonality property of the Legendre polynomials

$$\int_0^{\pi} P_m(\cos\phi) P_n(\cos\phi) \sin\phi \, d\phi = \int_0^1 P_m(x) P_n(x) \, dx = \begin{cases} 0 & m \ne n \\ \dfrac{2}{2m+1} & m = n \end{cases}$$

we have

$$\left(\frac{2}{2m+1}\right) a_m = \int_0^{\pi} u(1,\phi) P_m(\cos\phi) \sin\phi \, d\phi$$

$$= \int_0^{\pi/2} P_m(\cos\phi) \sin\phi \, d\phi - \int_{\pi/2}^{\pi} P_m(\cos\phi) \sin\phi \, d\phi$$

$$= \begin{cases} 0 & m = 0, 2, 4, \ldots \\ 2\int_0^1 P_m(x) \, dx & m = 1, 3, 5, \ldots \end{cases}$$

Solving for a_m, we find $a_m = 0$, $m = 0, 2, 4, \ldots$ and

$$a_m = (2m+1)\int_0^1 P_m(x)\, dx = (2m+1)\left[\frac{\sqrt{\pi}}{2\Gamma\left(1-\dfrac{m}{2}\right)\Gamma\left(\dfrac{m+3}{2}\right)}\right],\ m = 1,3,\ldots$$

where $\Gamma(\ )$ is the gamma function. Although we could evaluate the gamma function on Mathematica or even at wolframalpha.com, the above integral is best left for Mathematica or Maple to evaluate numerically, and doing this using Mathematica commands

$$3*\text{NIntegrate}[\text{LegendreP}[1,x], \{x,0,1\}]$$
$$5*\text{NIntegrate}[\text{LegendreP}[3,x], \{x,0,1\}]$$

gives $a_1 = 3/2$, $a_3 = -7/8$ and so the first few terms of the solution are

$$u(r,\phi) = \sum_{n=0}^{\infty} a_n r^n P_n(\cos\phi)$$

$$\doteq \frac{3}{2} r P_1(\cos\phi) - \frac{7}{8} r^3 P_3(\cos\phi)$$

$$= \frac{3}{2} r \cos\phi - \frac{7}{8} r^3 \left[\frac{5\cos^3\phi - 3\cos\phi}{2}\right]$$

$$= \left[\frac{21}{16} r^3 + \frac{3}{2} r\right]\cos\phi - \frac{35}{16} r^3 \cos^3\phi$$

Note that this solution says that

$$u(1,0) = \frac{5}{8}, u(1,\pi/2) = 0, u(1,\pi) = -\frac{5}{8}$$

which means the temperature on the surface of the sphere will be zero at the equator $\phi = \pi/2$, 5/8 at the north pole, and -5/8 at the south pole. You can play around with this solution by yourself going from the center of the sphere to the boundary (r going from 0 to 1) at $\phi = \pi/4$ (45 degrees north latitude) to see how the temperature changes. In fact, we will do it getting the curve

Lesson 35: Laplace's Equation in Spherical Coordinates

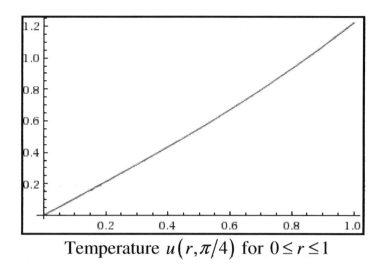

Temperature $u(r, \pi/4)$ for $0 \leq r \leq 1$

6. Solve the exterior Dirichlet problem

PDE: $(r^2 u_r)_r + \dfrac{1}{\sin\phi}\left[\sin\phi\, u_\phi\right]_\phi + \dfrac{1}{\sin^2\phi} u_{\theta\theta} = 0 \quad 1 < r < \infty$

BC: $u(1,\phi) = 1 + \cos\phi \quad 0 \leq \phi < \pi$

Solution: Since the BC is independent of θ, the solution has the form

$$u(r,\phi) = \sum_{n=0}^{\infty} \dfrac{b_n}{r^{n+1}} P_n(\cos\phi)$$

We first find the Legendre polynomials

$$P_0(\cos\phi) = 1$$
$$P_1(\cos\phi) = \cos\phi$$
$$P_2(\cos\phi) = \dfrac{1}{2}\left[3\cos^2\phi - 1\right]$$
$$P_3(\cos\phi) = \dfrac{1}{2}\left[5\cos^3\phi - 3\cos\phi\right]$$
$$\ldots \quad \ldots \quad \ldots$$

Plugging the general form of the solution in the BC we have

$$u(1,\phi) = \sum_{n=0}^{\infty} b_n P_n(\cos\phi) = 1 + \cos\phi$$

which requires

$$b_n = \begin{cases} 1 & n=0 \\ 1 & n=1 \\ 0 & n=2,3,\ldots \end{cases}$$

Hence, the solution

$$u(r,\phi) = \sum_{n=0}^{\infty} \frac{b_n}{r^{n+1}} P_n(\cos\phi)$$

$$= P_0(\cos\phi) + \frac{1}{r} P(\cos\phi)$$

$$= \frac{1}{r} + \frac{1}{r^2}\cos\phi$$

ΣΞΨΩZΥ

Lesson 36: A Nonhomogeneous Dirichlet Problem (Green's Function)

> 1. Find the potential due to a point source in three dimensions.

Solution The total outward flux across the boundary of the sphere of radius r is

$$\text{outward flux} = -4\pi r^2 u_r$$

and since the total outward flux is equal to the total charge inside the sphere, we set this value equal to q, getting

$$-4\pi r^2 u_r = q \Rightarrow u_r = -\frac{q}{4\pi r^2}$$

and solving for $u(r)$ yields

$$u(r) = \frac{q}{4\pi r}$$

> 2. Find Green's function $G(x, y, \xi, \eta)$ for Laplace's equation in the upper half plane $y \geq 0$. In other words, find the potential in the upper half plane at the point (x, y) (zero on the boundary $y = 0$) due to a point charge at (ξ, η). Hint: If we place a negative charge at $\overline{Q} = (\xi, -\eta)$ then it is clear that the potential field on the line $y = 0$ due to the two charges at $Q = (\xi, \eta)$ and $\overline{Q} = (\xi, -\eta)$ is zero. Hence, Green's function would be the result due to these two charges.

Solution: Using the strategy hinted at in the problem, we place a negative charge at $\overline{Q} = (\xi, -\eta)$ which along with the positive charge at $Q = (\xi, \eta)$ results in a zero charge on the x-axis. Adding the logarithmic potential due to the positive charge at $Q = (\xi, \eta)$ and the negative charge at $\overline{Q} = (\xi, -\eta)$, the charge at (x, y), or Green's function, will be

Lesson 36: A Nonhomogeneous Dirichlet Problem

$$G(x,y,\xi,\eta) = \frac{1}{2\pi}\ln(R) - \frac{1}{2\pi}\ln(\bar{R})$$
$$= \frac{1}{2\pi}\ln\left(\frac{R}{\bar{R}}\right)$$

which satisfies Laplace's equation for all (x,y) in the upper half plane except at $Q=(\xi,\eta)$, and $R, \bar{R}$ are the distances from (x,y) to (ξ,η) and $(\xi,-\eta)$, respectively, i.e.

$$R = \sqrt{(x-\xi)^2 + (y-\eta)^2}$$
$$\bar{R} = \sqrt{(x-\xi)^2 + (y+\eta)^2}$$

Note: Observe that the Green's function is zero for $y=0$ as desired since

$$G(x,0,\xi,\eta) = \frac{1}{2\pi}\ln(R) - \frac{1}{2\pi}\ln(\bar{R})$$
$$= \frac{1}{4\pi}\left\{\ln\left[(x-\xi)^2 + (-\eta)^2\right] - \ln\left[(x-\xi)^2 + (\eta)^2\right]\right\} = 0$$

---

3. Use the results of Problem 2 to find the solution $u(x,y)$ of Poisson's equation

   PDE: $u_{xx} + u_{yy} = -\rho \quad -\infty < x < \infty, \ 0 < y < \infty$

   BC: $u(x,0) = 0 \quad -\infty < x < \infty$

Solution: Green's form of the solution of the Dirichlet problem

PDE: $u_{xx} + u_{yy} = f(x,y) \quad -\infty < x < \infty, \ 0 < y < \infty$

BC: $u(x,0) = 0 \quad -\infty < x < \infty$

is

$$u(x,y) = \int_0^\infty \int_{-\infty}^\infty G(x,y,\xi,\eta) f(\xi,\eta) d\xi d\eta$$

We saw in Problem 2 that

Lesson 36: A Nonhomogeneous Dirichlet Problem

$$G(x,y,\xi,\eta) = \frac{1}{2\pi}\ln\left(\frac{R}{\overline{R}}\right) = \frac{1}{4\pi}\ln\left(\frac{(x-\xi)^2+(y-\eta)^2}{(x-\xi)^2+(y+\eta)^2}\right)$$

where

$$R = \sqrt{(x-\xi)^2+(y-\eta)^2}$$

$$\overline{R} = \sqrt{(x-\xi)^2+(y+\eta)^2}$$

Hence, the solution with $f(x,y) = -\rho$ is

$$u(x,y) = -\frac{\rho}{4\pi}\int_0^\infty\int_{-\infty}^\infty \ln\left(\frac{(x-\xi)^2+(y-\eta)^2}{(x-\xi)^2+(y+\eta)^2}\right) d\xi\, d\eta$$

This integral will give on the potential at a point (x,y) in the upper half plane. However, there are problems with this integral for two reasons. First, the integrand is infinite when $(x,y)=(\xi,\eta)$ and secondly, the domain is infinite. What we can do however to find $u(x,y)$ is not to integrate on a small circle around (x,y) and take smaller and smaller circles getting the limiting value, which will exist. Seconding do not integrate over the entire half plane but on a large rectangle, taking larger and larger rectangles to determining the limiting value. In other words, this integral formula is generally useful for finding the potential $u(x,y)$ at a few points (x,y). If one wants to find $u(x,y)$ over a wide range of values, it would be better to construct a grid and use some numerical method like we will study shortly.

---

4. How would you construct Green's function for the first quadrant $x \geq 0$, $y \geq 0$?

Solution: Place positive charges at (ξ,η) and $(-\xi,-\eta)$ and negative charges $(-\xi,\eta)$ and $(\xi,-\eta)$ so Green's function will be

$$G(x,y,\xi,\eta) = \frac{1}{2\pi}\ln(R_1) + \frac{1}{2\pi}\ln(R_3) - \frac{1}{2\pi}\ln(R_2) - \frac{1}{2\pi}\ln(R_4)$$
$$= \frac{1}{2\pi}\ln\left(\frac{R_1 R_3}{R_2 R_4}\right)$$

where $R_1, R_2, R_3,$ and R_4 are the distances as illustrated in Figure 36.1.

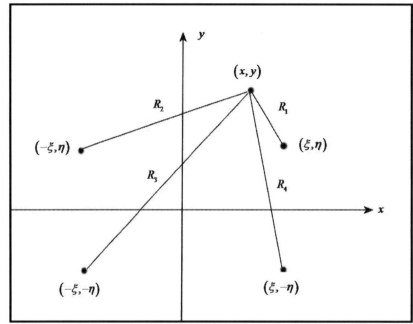

Figure 36.1 Positive charges at $(\xi,\eta),(-\xi,-\eta)$, negatives at $(\xi,-\eta),(-\xi,\eta)$

5. An alternative approach to solving Poisson's equation that sometimes works is the following. Suppose you want to find the solution of

$$\text{PDE: } u_{rr} + \frac{1}{r}u_r + \frac{1}{r^2}u_{\theta\theta} = 1 \quad 0 < r < 1$$
$$\text{BC: } u(1,\theta) = 1 \quad 0 \le \theta < 2\pi$$

You start by finding any solution of

$$u_{rr} + \frac{1}{r}u_r + \frac{1}{r^2}u_{\theta\theta} = 1$$

Lesson 36: A Nonhomogeneous Dirichlet Problem

> by trying a solution of the form $U(r,\theta) = Ar^2$ into the PDE and solving for A. After finding this solution, let $u = W + U$.

Solution: Trying

$$u_p(r) = Ar^2$$

and substituting this quantity into the PDE, we find

$$u_p(r) = \frac{1}{4}r^2$$

If we now let

$$u = w + \frac{1}{4}r^2$$

we see that w satisfies the BVP with homogeneous PDE

$$\text{PDE: } u_{rr} + \frac{1}{r}u_r + \frac{1}{r^2}u_{\theta\theta} = 0 \quad 0 < r < 1$$

$$\text{BC: } u(1,\theta) = -\frac{1}{4} + \sin\theta \quad 0 \leq \theta < 2\pi$$

which has the solution

$$w(r,\theta) = -\frac{1}{4} + r\sin\theta$$

Hence

$$u(r,\theta) = w(r,\theta) + \frac{1}{4}r^2 = \frac{1}{4}(r^2 - 1) + r\sin\theta$$

$$\Sigma K \Psi Z \Upsilon \Omega$$

The God Equation: Sentry at the Pearly Gates

"Wrong!" the old woman at the gate cackled fiendishly in a voice that could raise the dead, which considering the fact that everyone within earshot *was* dead, made the entire line jump to attention. The meek-looking man who was the object of her ridicule lowered his head and slunk off.

"Next," the old bag hissed. I had finally reached the front of the line, whereupon I approached the gatekeeper and gave her my manila envelope. She tore it open and began rifling through the papers. "Farlow?" she said looking over a pair of antiquated spectacles. "Is that *you*, little Jerry Farlow?

I looked into the beady eyes that peered from the face of this sorceress. Suddenly something inside clicked.

"Ms Hammerschnozle?" I asked. "Is that you?"

"Of course it's me. *But what are you doing here?*"

"We all die sometime, Ms. Hammerschnozle," I said. "I didn't want to go to the other place. But what are you doing here?"

"*Where did you think I'd be?*" she said snidely.

"Oh, I knew you'd be here," I lied through my teeth.

"But isn't Saint Peter supposed to be a *male* angel."

"Well, as usual Mr. Farlow, you thought wrong," The old bag said. "Times have changed. They finally got some religion up here and brought in someone that knew a thing or two about math."

God! My ultimate worst nightmare! My infamous tenth-grade math teacher, Ms. Hammerschnozle, whose endless aspersions on my academic reputation almost a century before, was now a stand-in at the Pearly Gates for no other than, alas, Saint Peter. And she would determine whether my final destination would be the `ol harp farm in the sky or a permanent reservation along the River Styx.

"You know," I smiled at Ms. Hammerschnozle. "I've learned just a little mathematics myself since my school days. Why don't you just give me the test." I smiled smugly, knowing there was no way she could ask me something about mathematics I didn't know.

"Ok, Farlow," she said. "What do you know about the *God Equation*."

"*The God Equation?*" I croaked. A river of sweat the size of the Ganges ran down my face. "Uh, I don't think I'm familiar with that particular equation."

"It's God's own equation," she said. "This single equation contains the most famous numbers in all mathematics. If you can answer my questions about these numbers, you will be allowed to enter Heaven. If not you will ... ."

"Yes, yes, I understand," I shuddered to think about the alternative.

"Are you ready, Farlow?" she asked. "The first number is about the number 1. What do you know about 1 ?"

The Number 1:

"Was she kidding?" I chuckled to myself. *Eins, Zwei* good `ol Numero Uno, the first of what we call the *natural numbers*

$$1, 2, 3, 4, ...$$

In addition to being the first natural number, it is the only number that has the property that the product is unchanged when multiplied by other numbers. For example

$$1 \times 6 = 6 \quad 1 \times 34 = 3 \quad 1 \times 68 = 68$$

"Ok," Hammerschnozle interrupted, "Enough of the kid stuff. I think we can get on to something more challenging." I waited for the guillotine to drop. "What can you tell me about π, the most famous number in geometry?"

The Number π:

I chuckled to myself. Although π isn't an integer like 1, it is one of the most famous numbers in mathematics and was well known by Greek mathematicians over two thousand years ago. Quite simply, it is the ratio of the circumference of a circle to its diameter. The first eight digits of π are 3.1415926, which means that the circumference of a circle is slightly more than three times its diameter. It is also ... ."

"Ok Farlow, you can stop with all that," Hammerschnozle interrupted. "You're halfway home. Are you ready for the third great number of mathematics?" she asked. "Each number gets just a bit harder."

I started to squirm a little. "Yes," I finally said.

"Ok," she said. "What can you tell me about the imaginary number *i*?"

The Imaginary Number *i* :

I was now on a roll. The imaginary number *i* is one of the most fascinating numbers in all mathematics. You might say the origin of the number *i* began in the 16th century when Italian mathematicians Cardano and Tartaglia tried to solve 3rd and 4th order polynomial equations. In the process they ran into the square root of -1, or $\sqrt{-1}$. They considered the square root of a negative number an impossibility so it was taken as a 'non number,' but after some time mathematicians found it useful to call $\sqrt{-1}$ a number, sometimes called an imaginary number, denoted by *i*. The entire subject of complex numbers has applications in ...

"Ok, ok, you can stop now," Hammerschnozle interrupted. "It's clear you know a little about mathematics. You told me about the three famous numbers 1, π, and *i*. We now have only one more number to go but it is the most difficult. What can you tell me about the number *e*?"

The Elusive Number e:

"What?" I thought, no trick question? Every mathematician worth his salt knows about e. In some areas of mathematics, like differential equations, it is possibly the most famous number of all. The number e is a real number like the numbers 1 and π. However, it is more difficult to describe since it is generally defined as the limiting value of some other numbers. The Swiss mathematician, Leonhard Euler, defined the number e as the limiting value of the expression

$$\boxed{\left(1+\frac{1}{n}\right)^n}$$

The God Equation: Sentry at the Pearly Gates

as *n* gets larger and larger. It is used by engineers to describe growth and decay in ...

"That's enough," Hammerschnozle said at last, "No more sugar coating the questions."

I looked over her shoulder to see if the Pearly Gates had started to open. "Now just tell me about the God equation and we're finished," she said.

"*What?*" I said.

"Suppose we combine 1, π, *i* and *e* into the single quantity

$$\boxed{e^{i\pi}+1}$$

she said. "What new number do we get?"

"*It's impossible,*" I told her. Just the thought of combining these four numbers into the single complicated number $e^{i\pi}+1$ gave me chills.

"Wrong!" she barked. The value of $e^{i\pi}+1$ can easily be found. *In fact it's zero!* The great Swiss mathematician Leonhard Euler used his famous equation

$$\boxed{e^{ix} = \cos x + i \sin x}$$

and simply plugging in $x = \pi$, he got

$$\boxed{e^{i\pi} = \cos \pi + i \sin \pi = -1 + i(0) = -1}$$

thus

$$\boxed{e^{i\pi} + 1 = -1 + 1 = 0}$$

"Up here we call it the *God Equation*," Hammerschnozle said.

"*What?*" I gasped in disbelief.

She continued, "This single equation brings together the five most famous numbers in all mathematics; the basic whole numbers 0 and 1, the fundamental constant of geometry π, the basic complex number *i*, and the core constant of calculus, *e.*"

Next!" the old bag cackled heinously.

"*Aaaaaaaaaaaahhhhhhhhhhhhhhhhh,*" I screamed as two burly angels grabbed me and flew off. "No, no," I protested. "I belong here. Give me another chance."

"*Wake up, wake up,*" I felt someone tugging on my arm. I was sitting up in bed, drenched in sweat, looking at my wife. "Those nasty little monkeys carrying you off again, Toto?" she said.

"$e^{i\pi}+1=0$" I gasped. "Would you believe it? Did you know

$$\boxed{e^{i\pi}+1=0 \; ?}$$"

"Yeah, yeah" she said rolling over, "The God Equation. Go back to sleep."

Section 5: Numerical and Approximate Solutions

Lesson 37: Numeric Solution (Elliptic Problems)

1. Derive the following approximation equation for the second derivative.

$$f''(x) \cong \frac{1}{h^2}\left[f(x+h) - 2f(x) + f(x-h)\right]$$

Solution: Writing the first three terms of the Taylor series for $f(x+h)$ and $f(x-h)$ we have

$$f(x+h) \doteq f(x) + hf'(x) + \frac{h^2}{2}f''(x)$$

$$f(x-h) \doteq f(x) - hf'(x) + \frac{h^2}{2}f''(x)$$

Adding these equations, then solving for $f''(x)$, gives the desired result.

2. Carry out the computation for two iterations in the following Dirichlet problem using the Liebmann iterative process.

$$\text{PDE: } u_{xx} + u_{yy} = 0 \quad 0 < x < 1, 0 < y < 1$$

$$\text{BC: } u = \begin{cases} 0 \text{ on the top and sides of the square} \\ \sin(\pi x) \quad 0 \leq x \leq 1 \end{cases}$$

Lesson 37: Numeric Solution (Elliptic Problems)

Solution: Start by making an $n \times n$ grid in the unit square

$$S = \{(x,y): 0 \leq x \leq 1,\ 0 \leq y \leq 1\}$$

At the grid points on the top and sides set $u_{i,j} = 0$ and on the top, evaluate them at the appropriate value of $\sin(\pi x)$. Then initialize the value of $u_{i,j}$ at all the interior grid points to the average value of all $u_{i,j}$ on the boundary. After then evaluate

$$u_{i,j} = \frac{1}{4}\left(u_{i-1,j} + u_{i+1,j} + u_{i,j-1} + u_{i,j+1}\right),\ i,j = 1,\ldots,n$$

Then repeat this process again until you get a desired convergence of the interior values of $u_{i,j}$. You will discover that the numbers $u_{i,j}$ after a few iterations. But realize that they are not converging to the solution of the BV problem but to the solution of the difference equation, which is close to the solution of the BV problem. If you want a better approximation you should choose the grid size h smaller. A good strategy would be to pick a couple different grid sizes and see if you procedure converges to the same numbers.

---

3. What are the algebraic equations that must be solved when you use a finite-difference approximation to solve the following Poisson equation inside a square.

PDE: $u_{xx} + u_{yy} = f(x,t)\quad 0 < x < 1,\ 0 < y < 1$

BC: $u(x,y) = g(x,y)$ on the boundary

Solution: Using the finite difference approximations for the derivatives, we have

$$u_{xx} \doteq \frac{1}{h^2}\left(u_{i,j+1} - 2u_{i,j} + u_{i,j-1}\right) \quad i = 2,\ldots,n-1,\ j = 2,\ldots,n-1$$

$$u_{yy} \doteq \frac{1}{h^2}\left(u_{i+1,j} - 2u_{i,j} + u_{i-1,j}\right) \quad i = 2,\ldots,n-1,\ j = 2,\ldots,n-1$$

Plugging this into the PDE, we have

Lesson 37: Numeric Solution (Elliptic Problems)

$$\frac{1}{h^2}\left(u_{i,j+1} - 2u_{i,j} + u_{i,j-1}\right) + \frac{1}{h^2}\left(u_{i+1,j} - 2u_{i,j} + u_{i-1,j}\right) = f_{i,j}$$

and solving for $u_{i,j}$ gives

$$u_{i,j} = \frac{1}{4}\left[u_{i+1,j} + u_{i-1,j} + u_{i,j+1} + u_{i,j-1}\right] - \frac{h^2}{4}f_{i,j}$$

When the grid point (i, j) is adjacent to a boundary (i.e. if any of $i=2, i=n-1, j=2, j=n-1$, then the corresponding value of $g_{i,j}$ should replace the corresponding value of $u_{i,j}$.

---

4. What are the algebraic equations that must be solved when you use a finite-difference approximation to solve the following Poisson equation inside a square.

$$\text{PDE:} \quad u_{xx} + u_{yy} + 2u = 0 \quad 0 < x < 1, \, 0 < y < 1$$

$$\text{BC:} \quad u(x,y) = g(x,y) \text{ on the boundary}$$

Solution: Using the finite difference approximations for the derivatives, we have

$$u_{xx} \doteq \frac{1}{h^2}\left(u_{i,j+1} - 2u_{i,j} + u_{i,j-1}\right), \quad i=2,\ldots,n-1, \; j=2,\ldots,n-1$$

$$u_{yy} \doteq \frac{1}{h^2}\left(u_{i+1,j} - 2u_{i,j} + u_{i-1,j}\right), \quad i=2,\ldots,n-1, \; j=2,\ldots,n-1$$

Plugging this into the PDE, we have

$$\frac{1}{h^2}\left(u_{i,j+1} - 2u_{i,j} + u_{i,j-1}\right) + \frac{1}{h^2}\left(u_{i+1,j} - 2u_{i,j} + u_{i-1,j}\right) + 2u_{i,j} = 0$$

and solving for $u_{i,j}$ gives

$$u_{i,j} = \frac{-1}{2(h^2 - 2)}\left(u_{i,j+1} + u_{i,j-1} u_{i+1,j} + u_{i-1,j}\right)$$

Lesson 37: Numeric Solution (Elliptic Problems)

When the grid point (i, j) is adjacent to a boundary (i.e. if any of $i = 2, i = n-1, j = 2, j = n-1$, then the corresponding value of $g_{i,j}$ should replace the corresponding value of $u_{i,j}$.

---

5. How would you solve the following Neumann problem inside the square by the finite-difference method.

PDE: $u_{xx} + u_{yy} = 0$ $0 < x < 1, 0 < y < 1$

BCs: $\begin{cases} u(x, y) = 0 \text{ on the top, bottom, and left hand side} \\ \dfrac{\partial u}{\partial x}(1, y) = g(y) \quad 0 \leq y \leq 1 \end{cases}$

Solution: The finite difference formula for the PDE is

$$u_{i,j} = \frac{1}{4}\left[u_{i+1,j} + u_{i-1,j} + u_{i,j+1} + u_{i,j-1} \right]$$

for the interior grid points, and for the boundary points, we have

$u_{1,j} = 0$ (bottom side)

$u_{n,j} = 0$ (top side)

$u_{i,1} = 0$ (left side)

$u_{i,n} = u_{i,n-1} + hg(y_i)$ (right side)

268 Lesson 37: Numeric Solution (Elliptic Problems)

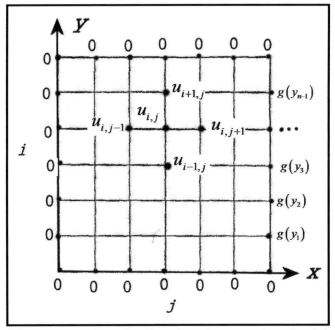

Typical grid construction for Laplace's equation inside a square.

For grid points $(i, n-1)$, $i = 1, 2, \ldots, n$ on the vertical line *next* to the right boundary grid points (i, n), $i = 1, 2, \ldots, n$, we use the formula $u_{i,n} = u_{i,n-1} + hg(y_i)$ to estimate the values of $u(1, y_i)$, $i = 1, 2, \ldots, n$, where h is the grid size. This will satisfy the derivative BC $u_x(1, y) = g(y)$, $0 \le y \le 1$ at the right side of the square.

---

6. Write a flow diagram to solve the following Dirichlet problem inside a square with an arbitrary number of grid lines.

$$\text{PDE: } u_{xx} + u_{yy} = f(x, y) \quad 0 < x < 1, 0 < y < 1$$

$$\text{BC: } u(x, y) = g(x, y) \text{ on the boundary}$$

Solution: We saw in Problem 3 the finite difference approximations for the derivatives are

$$u_{xx} \doteq \frac{1}{h^2}\left(u_{i,j+1} - 2u_{i,j} + u_{i,j-1}\right)$$

$$u_{yy} \doteq \frac{1}{h^2}\left(u_{i+1,j} - 2u_{i,j} + u_{i-1,j}\right)$$

Lesson 37: Numeric Solution (Elliptic Problems)

Plugging this into the PDE, we have

$$\frac{1}{h^2}\left(u_{i,j+1} - 2u_{i,j} + u_{i,j-1}\right) + \frac{1}{h^2}\left(u_{i+1,j} - 2u_{i,j} + u_{i-1,j}\right) = f_{i,j}$$

and solving for $u_{i,j}$ gives

$$u_{i,j} = \frac{1}{4}\left[u_{i+1,j} + u_{i-1,j} + u_{i,j+1} + u_{i,j-1}\right] - \frac{h^2}{4} f_{i,j}$$

This is the equation that is used to approximate the solution of the problem. We leave the details of a flow diagram for the reader.

$$\Sigma \Upsilon \Omega \Xi \Psi \Theta$$

Lesson 38: An Explicit Finite-Difference Method

> 1. Find the finite-difference approximation of the following heat conduction problem for $t = 0.005, 0.010, 0.0015$ by the explicit method. Let $h = \Delta x = 0.1$. Plot the solution at $x = 0, 0.1, 0.2, 0.3, 0.4,...,0.9,1$ for $t = 0.015$.
>
> $$\text{PDE: } u_t = u_{xx} \quad 0 < x < 1,\ 0 < t < \infty$$
>
> $$\text{BCs: } \begin{cases} u(0,t) = 0 \\ u(1,t) = 0 \end{cases} \quad 0 < t < \infty$$
>
> $$\text{IC: } u(x,0) = \sin(\pi x) \quad 0 \leq x \leq 1$$

Solution: Student project

> 2. Solve
>
> $$\text{PDE: } u_t = u_{xx} \quad 0 < x < 1,\ 0 < t < \infty$$
>
> $$\text{BCs: } \begin{cases} u(0,t) = 0 \\ u(1,t) = 0 \end{cases} \quad 0 < t < \infty$$
>
> $$\text{IC: } u(x,0) = \sin(\pi x) \quad 0 \leq x \leq 1$$
>
> by separation of variables, then evaluate the analytical solution at the same grid points where you carried out the finite difference approximation in Problem 1.

Solution: We find the analytic solution by seeking product solutons

$$u(x,t) = X(x)T(t)$$

Plugging this equation in the PDE gives the two ODEs

$$\begin{aligned} T' + (n\pi)^2 T &= 0 \\ X'' + (n\pi)^2 X &= 0 \end{aligned} \quad n = 1, 2, ...$$

Lesson 38: An Explicit Finite-Difference Method

which has solutions

$$T_n(t) = c_n e^{-(n\pi)^2 t}$$
$$X_n(x) = a_n \cos(n\theta) + b_n \sin(n\theta)$$

$n = 1, 2, ...$

By superposition we have the series solution

$$u(x,t) = \sum_{n=1}^{\infty} T_n(t) X_n(x) = \sum_{n=1}^{\infty} e^{-(n\pi)^2 t} \left[A_n \cos(n\theta) + B_n \sin(n\theta) \right]$$

The last step is to find the coefficients A_n, B_n so that $u(x,t)$ satisfies the IC

$$u(x,0) = \sin(\pi x) \quad 0 \leq x \leq 1$$

It is easy to see that all coefficients must be zero, except $B_1 = 1$. Hence, we have the solution

$$u(x,t) = e^{-\pi^2 t} \sin(\pi x), \ 0 \leq x \leq 1, \ 0 \leq t < \infty$$

We leave it to the reader to compare the numeric solutions found in Problem 1 with these values.

---

3. Write a flow diagram to carry out the computations of the following hyperbolic problem.

PDE: $u_{tt} = u_{xx}$ $0 < x < 1, \ 0 < t < \infty$

BCs: $\begin{cases} u(0,t) = g_1(t) \\ u(1,t) = g_2(t) \end{cases}$ $0 < t < \infty$

ICs: $\begin{cases} u(x,0) = \phi(x) \\ u_t(x,0) = \psi(x) \end{cases}$ $0 \leq x \leq 1$

Solution: Student project

4. Find the finite-difference approximation of the following heat conduction problem for $t = 0.005, 0.010, 0.0015$ by the explicit method. Let $h = \Delta x = 0.1$. Plot the solution at $x = 0, 0.1, 0.2, 0.3, 0.4, ..., 0.9, 1$ for $t = 0.015$.

$$\text{PDE: } u_t = u_{xx} \quad 0 < x < 1, \ 0 < t < \infty$$

$$\text{BCs: } \begin{cases} u(0,t) = 0 \\ u_x(1,t) = -[u(1,t) - 1] \end{cases} \quad 0 < t < \infty$$

$$\text{IC: } u(x,0) = \sin(\pi x) \quad 0 \leq x \leq 1$$

Solution: Student project

ΣΥΡΞΨΩ

Lesson 39: An Implicit Finite-Difference Method (Crank-Nicolson Method)

1. Show that the implicit Crank Nicolson method applied to the IBVP

$$\text{PDE: } u_t = u_{xx} \quad 0 < x < 1, \; 0 < t < \infty$$

$$\text{BC: } \begin{cases} u(0,t) = 0 \\ u(1,t) = 0 \end{cases} \quad 0 < t < \infty$$

$$\text{IC: } u(x,0) = 0 \quad 0 \le x \le 1$$

gives rise to the formula

$$-\lambda r u_{i+1,j+1} + (1+2r\lambda) u_{i+1,j} - \lambda r u_{i+1,j-1} = r(1-\lambda) u_{i,j+1} + [1 - 2r(1-\lambda)] u_{i,j} + r(1-\lambda) u_{i,j-1}$$

Solution: Direct computation by plugging in the proper finite difference formulas for u_t and u_{xx} as illustrated in the book. The BC and IC should be replaced by

$$\text{BCs: } \begin{cases} u_{i,1} = 0 \\ u_{i,n} = 0 \end{cases} \quad i = 1, 2, \ldots, m$$

$$\text{IC: } u_{i,j} = 1 \quad j = 1, 2, \ldots, n-1$$

2. How would you solve the following problem by the implicit finite-difference method.

$$\text{PDE: } u_t = u_{xx} - u_x \quad 0 < x < 1, \; 0 < t < \infty$$

$$\text{BCs: } \begin{cases} u(0,t) = 1 \\ u_x(1,t) + u(1,t) = g(t) \end{cases} \quad 0 < t < \infty$$

$$\text{ICs: } u(x,0) = 0 \quad 0 \le x \le 1$$

Solution: The finite difference approximation for the PDE is the same as in the lesson. The finite difference approximations to the BCs should be

Lesson 39: An Implicit Finite-Difference Method

$$u_{i,1} = 1$$

$$\frac{u_{i,n} - u_{i,n-1}}{h} + u_{i,n} = g_i$$

where $g_i = g(ik)$. For each value of time we solve $n-1$ equations for

$$u_{i,2}, u_{i,3}, u_{i,4}, \ldots, u_{i,n}$$

The BC at $x=1$ will give you an extra equation.

---

3. How would you solve the following problem by the implicit finite-difference method.

PDE: $u_t = u_{xx} + u \quad 0 < x < 1, \ 0 < t < \infty$

BCs: $\begin{cases} u(0,t) = 1 \\ u(1,t) = 0 \end{cases} \quad 0 < t < \infty$

ICs: $u(x,0) = 1 \quad 0 \le x \le 1$

Solution: This problem is almost identical to the one in the lesson. We let you carry out the computations.

---

4. What is the molecular form of the following equation if we pick $\lambda = 1$?

$$-\lambda r u_{i+1,j+1} + (1+2r\lambda) u_{i+1,j} - \lambda r u_{i+1,j-1} = r(1-\lambda) u_{i,j+1} + [1-2r(1-\lambda)] u_{ij} + r(1-\lambda) u_{i,j-1}$$

Solution: The molecular form in drawn in Figure 39.1.

Lesson 39: An Implicit Finite-Difference Method

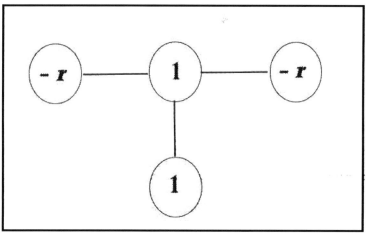

Figure 39.1 Molecular form of implicit method

5. Write a flow diagram to approximate the solution of the following heat problem.

PDE: $u_t = u_{xx}$ $0 < x < 1,\ 0 < t < \infty$

BCs: $\begin{cases} u(0,t) = 1 \\ u(1,t) = 0 \end{cases}$ $0 < t < \infty$

ICs: $u(x,0) = 1$ $0 \le x \le 1$

Solution: Direct application of the formula from the lesson.

6. Solve the following system of equations

$$\begin{bmatrix} 3 & -1 & 0 & 0 \\ -1 & 3 & -1 & 0 \\ 0 & -1 & 3 & -1 \\ 0 & 0 & -1 & 3 \end{bmatrix} \begin{bmatrix} u_{23} \\ u_{24} \\ u_{25} \\ u_{26} \end{bmatrix} = \begin{bmatrix} 1 \\ 1 \\ 1 \\ 1 \end{bmatrix}$$

Solution: Student project using the iterative method.

ΞΩΨΠΥΘ

Lesson 40: Analytic versus Numeric Solutions

---

1. How would you construct an experiment to estimate the parameter α in the problem

$$\text{PDE: } u_t = \alpha^2 u_{xx} \quad 0 < x < 1, \ 0 < t < \infty$$

$$\text{BCs: } \begin{cases} u(0,t) = 1 \\ u(1,t) = 0 \end{cases} \quad 0 < t < \infty$$

$$\text{ICs: } u(x,0) = 1 \quad 0 \le x \le 1$$

using the analytical solution

$$u(x,t) = \frac{2}{\pi}\left[e^{-(\pi\alpha)^2 t} \sin(\pi x) + \frac{1}{3} e^{-(3\pi\alpha)^2 t} \sin(3\pi x) + \cdots \right]$$

Solution: Writing the first two terms of the analytical solution gives

$$u(x,t) \doteq \frac{2}{\pi}\left[e^{-(\pi\alpha)^2 t} \sin(\pi x) + \frac{1}{3} e^{-(3\pi\alpha)^2 t} \sin(3\pi x) \right]$$

Hence, when $x = 0.5$ we have

$$u(0.5,t) \doteq \frac{2}{\pi}\left[e^{-(\pi\alpha)^2 t} \sin\left(\frac{\pi}{2}\right) + \frac{1}{3} e^{-(3\pi\alpha)^2 t} \sin\left(\frac{3\pi}{2}\right) \right]$$

$$= \frac{2}{\pi}\left[e^{-(\pi\alpha)^2 t} - \frac{1}{3} e^{-(3\pi\alpha)^2 t} \right]$$

We can now perform an experiment where we observe values of u at $x = 0.5$ for different values of time. We can then do a least squares analysis to find the value of α that best fits the above function to the data points.

Lesson 40: Analytic versus Numeric Solutions 277

2. Least-squares approximation minimizes the sum of squares (SS) between a curve and the data points. For example, the least squares line $y(x) = a + bx$ that approximates the data points $(x_1, y_1), (x_2, y_2), ..., (x_n, y_n)$, would be the line that minimizes

$$S(a,b) = \sum_{i=1}^{n} \left[y_i - (a + bx_i) \right]^2$$

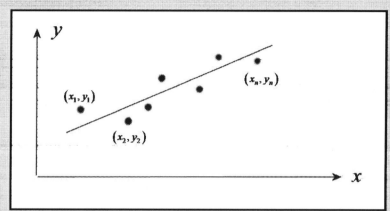

Figure 40.1 Least Squares diagram

Find the constants a, b in terms of the data points that minimizes SS.

Solution: We differentiate $S = S(a,b)$ which is a function of a and b with respect to a and b, and set the derivatives to zero, getting

$$\frac{\partial S}{\partial a} = 2 \sum_{i=1}^{n} \left[y_i - (a + bx_i) \right](-1) = 0$$

$$\frac{\partial S}{\partial b} = 2 \sum_{i=1}^{n} \left[y_i - (a + bx_i) \right](-x_i) = 0$$

or

$$a \sum_{i=1}^{n} 1 + b \sum_{i=1}^{n} x_i = \sum_{i=1}^{n} y_i$$

$$a \sum_{i=1}^{n} x_i + b \sum_{i=1}^{n} x_i^2 = \sum_{i=1}^{n} x_i y_i$$

or in matrix form, we have

$$\begin{bmatrix} n & \sum_{i=1}^{n} x_i \\ \sum_{i=1}^{n} x_i & \sum_{i=1}^{n} x_i^2 \end{bmatrix} \begin{bmatrix} a \\ b \end{bmatrix} = \begin{bmatrix} \sum_{i=1}^{n} y_i \\ \sum_{i=1}^{n} x_i y_i \end{bmatrix}$$

from which we can (easily but tedious) solve for a, b.

---

3. An important problem in biochemistry is the determination of the molecular weight of the macromolecule myoglobin. One approach is to place a certain blood solution in an ultracentrifuge and spin it for a given length of time. The equation that describes the concentration of the liquid in the centrifuge is known as Lamm's equation

$$\frac{\partial u(r,t)}{\partial t} = \frac{1}{r} \frac{\partial}{\partial r} \left(Dr \frac{\partial u}{\partial r} - s\omega^2 u \right)$$

where
   $r =$ distance from the center of the centrifuge
   $D =$ diffusion coefficient (depends on the weight of myoglobin)
   $s =$ sedimentation coefficient (computed experimentally)
   $w =$ angular velocity of the centrifuge (known)
   $u(r,t) =$ concentration of the medium in the centrifuge

The approach in finding the molecular weight of myoglobin is to find the steady-state solution $u(r, \infty)$ of Lamm's equation by letting $u_t = 0$ and solving the ODE (which is not difficult) and found to be

$$\frac{d}{dr}\left(Dr \frac{du}{dr} - s\omega^2 u(r) \right) = Dr \frac{d^2 u}{dr^2} + (D - s\omega^2) \frac{du}{dr} = 0$$

getting

$$U(r) = \left(\frac{1}{r} \right)^{(D - s\omega^2)/D} \quad 0 < r < \infty$$

How would you use this analytical solution to design an experiment to estimate the molecular weight α of myoglobin?

Solution: One designs an experiment where we measure the concentration of myoglobin at different values of r after the substance has come to steady state. We can then do a least squares fit of this date on the above analytical solution to estimate D, assuming we know s and ω.

$$\Sigma\Omega\Psi\Upsilon\Xi\Theta$$

Lesson 41: Classification of PDEs (Parabolic and Elliptic Equations)

> 1. Which of the following parabolic and elliptic equations are in canonical form?
>
> a) $u_t = u_{xx} - hu$
>
> b) $u_{xy} + u_{xx} + 3u = \sin x$
>
> c) $u_{xx} + 3u_{yy} = 0$
>
> d) $u_{xx} = \sin x$

Solution: The general second order, linear PDE in two independent variables is

$$Au_{xx} + Bu_{xy} + Cu_{yy} + Du_x + Eu_y + Fu = G$$

where A, B, C, D, E, F, G are functions (which of course includes constants) of x, y. If

$$B^2 - 4AC < 0 \Rightarrow \text{PDE is elliptic}$$
$$B^2 - 4AC = 0 \Rightarrow \text{PDE is parabolic}$$
$$B^2 - 4AC > 0 \Rightarrow \text{PDE is hyperbolic}$$

We also saw that the canonical form for each type of equation is

canonical forms for a hyperbolic equation	$\begin{cases} u_{\xi\xi} - u_{\eta\eta} = \Psi(\xi, \eta, u, u_\xi, u_\eta) \\ u_{\xi\eta} = \Phi(\xi, \eta, u, u_\xi, u_\eta) \end{cases}$
canonical parabolic equation	$u_{\eta\eta} = \Phi(\xi, \eta, u, u_\xi, u_\eta)$
canonical elliptic equation	$u_{\xi\xi} + u_{\eta\eta} = \Phi(\xi, \eta, u, u_\xi, u_\eta)$

Identifying these coefficients, we have the following classification

Lesson 41: Classification of PDEs (Parabolic and Elliptic Eq)

a) $u_t = u_{xx} - hu \Rightarrow B^2 - 4AC = 0 \Rightarrow$ parabolic, canonical

b) $u_{xy} + u_{xx} + 3u = \sin x \Rightarrow B^2 - 4AC = 1 > 0 \Rightarrow$ hyperbolic, not canonical

c) $u_{xx} + 3u_{yy} = 0 \Rightarrow B^2 - 4AC = -12 < 0 \Rightarrow$ ellliptic, not canonical

d) $u_{xx} = \sin x \Rightarrow B^2 - 4AC = 0 \Rightarrow$ parabolic, canonical

You should realize that the notation for the two independent variables, whether x, y or ξ, η is irrelevant.

> 2. Transform the following parabolic equation into canonical form.
>
> $$u_{xx} + 2u_{xy} + u_{yy} + u = 2$$

Solution: In this case we have

$$A = 1, B = 2, C = 1 \Rightarrow B^2 - 4AC = 0$$

and so there is only one characteristic equation

$$\frac{dy}{dx} = \frac{B - \sqrt{B^2 - 4AC}}{2A} = \frac{2 - \sqrt{2^2 - 4(1)(1)}}{2(1)} = 1$$

which gives the single characteristic equation

$$y = x + c_1$$

Solving for c_1 gives $c_1 = y - x$ and so $\xi = y - x$ (the idea is to pick the new coordinate on constant curves or lines in this case). The second coordinate η is arbitrary as long as the coordinate lines $\eta = $ constant are not parallel to those of $\xi = y - x = $ constant, so we choose something simple, like $\eta = y$, which give us the new coordinate system

$$\xi = y - x$$
$$\eta = y$$

We could have just as well chosen $\eta = x$, $\eta = x + y$, $\eta = y - 2x$,... but not $\eta = 2y - 2x$. We now use the diagram in Figure 41.1

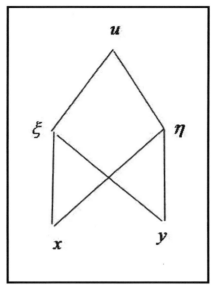

Figure 41.1 Relational diagram useful for computing derivatives

to assist in finding the following partial derivatives:

$$u_x = u_\xi \xi_x + u_\eta \eta_x = -u_\xi$$
$$u_y = u_\xi \xi_y + u_\eta \eta_y = u_\xi + u_\eta$$
$$u_{xx} = (u_x)_x = (-3u_\xi)_x = -(u_{\xi\xi}\xi_x + u_{\xi\eta}\eta_x) = u_{\xi\xi}$$
$$u_{yy} = (u_y)_y = (u_\xi + u_\eta)_y = u_{\xi\xi}\xi_y + u_{\xi\eta}\eta_y + u_{\xi\eta}\eta_y + u_{\eta\eta}\eta_y = u_{\xi\xi} + 2u_{\xi\eta} + u_{\eta\eta}$$
$$u_{xy} = (u_x)_y = -(u_\xi)_y = -(u_{\xi\xi}\xi_y + u_{\xi\eta}\eta_y) = -(u_{\xi\xi} + u_{\xi\eta})$$

then plugging them into the PDE

$$u_{xx} + 2u_{xy} + u_{yy} + u = 2$$

gives

$$(u_{\xi\xi}) + 2\left[-(u_{\xi\xi} + u_{\xi\eta})\right] + (u_{\xi\xi} + 2u_{\xi\eta} + u_{\eta\eta}) + u = 2$$

or

$$u_{\eta\eta} + u = 2$$

which is in canonical form $u_{\eta\eta} = \Phi(\xi, \eta, u, u_\xi, u_\eta)$ of a parabolic equation.

Lesson 41: Classification of PDEs (Parabolic and Elliptic Eq)

> 3. Transform the following elliptic equation into canonical form.
>
> $$u_{xx} + 2u_{yy} + x^2 u_x = e^{-x^2/2} \quad (x \neq 0)$$

Solution: In this case we have

$$A = 1, B = 0, C = 2 \Rightarrow B^2 - 4AC = -8 < 0$$

so the equation is elliptic (for $x \neq 0$) and the characteristic equations are

$$\frac{dy}{dx} = \frac{B - \sqrt{B^2 - 4AC}}{2A} = \frac{0 - \sqrt{0^2 - 4(1)(2)}}{2(1)} = \frac{0 - \sqrt{0^2 - 4(1)(2)}}{2(1)} = -i\sqrt{2}$$

$$\frac{dy}{dx} = \frac{B + \sqrt{B^2 - 4AC}}{2A} = \frac{B + \sqrt{B^2 - 4AC}}{2A} = \frac{0 + \sqrt{0^2 - 4(1)(2)}}{2(1)} = i\sqrt{2}$$

Solving these equations gives

$$y = -i\sqrt{2}\, x + c_1$$
$$y = i\sqrt{2}\, x + c_2$$

and solving for the constants c_1, c_2 gives

$$c_1 = y + i\sqrt{2}\, x$$
$$c_2 = y - i\sqrt{2}\, x$$

Hence, we have the new coordinates

$$\xi = y + i\sqrt{2}\, x$$
$$\eta = y - i\sqrt{2}\, x$$

or which we prefer, real coordinates

$$\alpha = \frac{\xi + \eta}{2} = y$$

$$\beta = \frac{\xi - \eta}{2i} = \sqrt{2}\, x$$

We now use the diagram in Figure 41.2

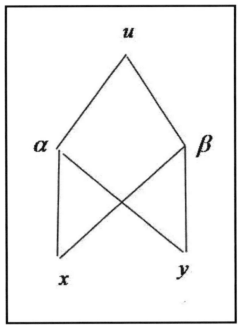

Figure 41.2 Functional diagram useful for computing partial derivatives

to assist in computing the derivatives

$$u_x = u_\alpha \alpha_x + u_\beta \beta_x = \sqrt{2}\, u_\beta$$
$$u_y = u_\alpha \alpha_y + u_\beta \beta_y = u_\alpha$$
$$u_{xx} = (u_x)_x = (\sqrt{2}\, u_\beta)_x = \sqrt{2}\left(u_{\alpha\beta}\alpha_x + u_{\beta\beta}\beta_x\right) = 2 u_{\beta\beta}$$
$$u_{yy} = (u_y)_y = (u_\alpha)_y = u_{\alpha\alpha}\alpha_y + u_{\alpha\beta}\beta_y = u_{\alpha\alpha}$$

and plugging them in the PDE

$$u_{xx} + 2 u_{yy} + x^2 u_x = e^{-x^2/2}$$

and using the fact that $x = \beta / \sqrt{2}$, we get the new PDE in ξ and η as

$$u_{\alpha\alpha} + u_{\beta\beta} + \frac{\sqrt{2}}{4}\beta^2 u_\beta = \frac{1}{2}e^{-\beta^2/4}$$

which is in canonical form $u_{\xi\xi} + u_{\eta\eta} = \Phi(\xi,\eta,u,u_\xi,u_\eta)$ of an elliptic equation. The notation of the independent variables α, β or ξ, η is irrelevant.

ΞΩΤΠΘΣ

Lesson 42: Monte Carlo Methods (an Introduction)

> 1. Write a computer program to estimate the value of the following integral and make a graph of the approximation as a function of the number of random tosses.
>
> $$I = \int_0^1 e^{\sin x}\, dx$$

Solution: The graph of the integrand $e^{\sin x}$ is drawn in Figure 42.1

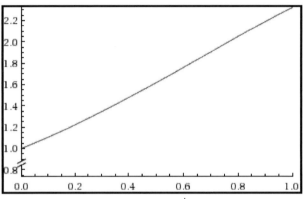

Figure 42.1 Graph of $e^{\sin x}$, $0 \le x \le 1$

The integrand is positive and has its value between 1 and $e^{\sin(1)} \doteq 2.32$, and so we generate a uniformly distributed random sample of points $(R_1, 2.32R_2)$, where R_1, R_2 are random numbers between 0 and 1, hence the points $(R_1, 2.32R_2)$ lie in the rectangle

$$R = \{(x, y): 0 \le x \le 1, 0 \le y \le 2.32\}$$

(like tossing darts randomly inside the rectangle) We then compute the fraction $\bar{f}$ of the points $(R_1, 2.32R_2)$ that that fall *below* the curve $y = e^{\sin x}$, from the rule

if $2.32R_2 < e^{\sin R_1}$ $\Rightarrow$ random point $(R_1, 2.32R_2)$ lies below the curve $e^{\sin x}$

if $2.32R_2 \ge e^{\sin R_1}$ $\Rightarrow$ random point $(R_1, 2.32R_2)$ lies on or above the curve $e^{\sin x}$

Lesson 42: Monte Carlo Methods (an Introduction)   287

We then multiply $\overline{f}$ by the *area* of the rectangle R, which is 2.32, which will be an approximation of the integral. That is

$$\int_0^1 e^{\sin x} \, dx \doteq 2.32 \, \overline{f}$$

If you carry out this simulation on a computer for say $n=100$ tosses, you will find roughly 70% of the tosses fall under the curve, hence $\overline{f} \doteq 0.70$, and multiplying this value by the area of the rectangle, which is 2.3, you will estimate the integral to be $0.7 \times 2.32 = 1.62$ (the correct value to 6 places is 1.63187). A flow diagram for this process is essentially like the flow diagram in Problem 3.

> 2. Generate a sequence of random numbers using the algorithm
>
> $$r_{n+1} \equiv (3r_n + 4) \bmod 7, \; r_0 = 0$$

Solution: The equation says:

1. take a random number r_n, the first one being $r_0 = 0$
2. compute $3r_n + 4$
3. divide by 7 and take the remainder, that will be r_{n+1}
4. repeat the process

Carrying out these computations, we will generate the sequence

$$0, 4, 2, 3, 6, 1, 0, \ldots \rightarrow \text{numbers repeat}$$

> 3. Write a flow diagram and computer program to estimate the following triple integral.
>
> $$I = \int_0^1 \int_0^1 \int_0^1 e^{-(x^2 + y + z^2)} \, dx \, dy \, dz$$

Solution:

Lesson 42: Monte Carlo Methods (an Introduction)

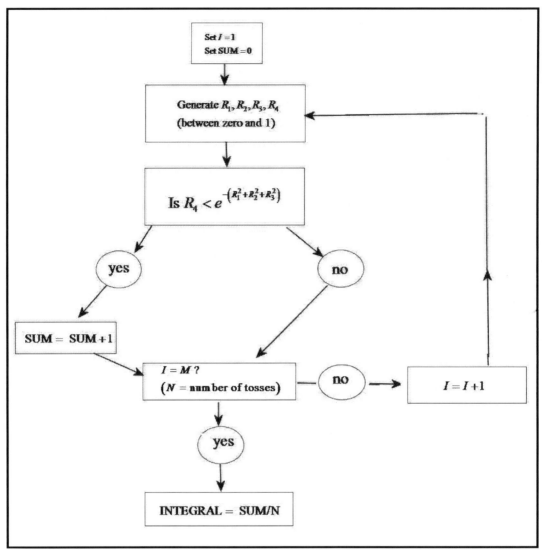

Figure 42.1 Flow diagram for approximating a triple integral

4. How would you generate a sequence of random points inside the following triangle?

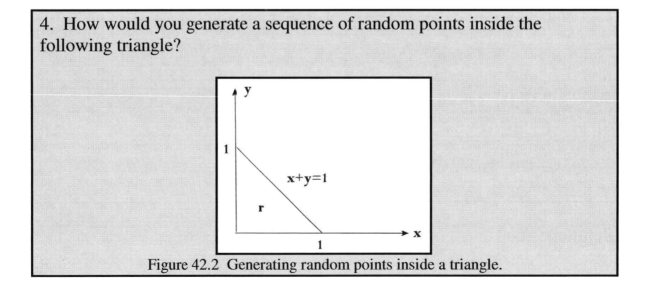

Figure 42.2 Generating random points inside a triangle.

Lesson 42: Monte Carlo Methods (an Introduction)

Solution: Generate random numbers R_1, R_2 between 0 and 1. If

$$R_1 + R_2 < 1$$

we keep the point, otherwise generate another pair and repeat the process. Continue this process to get a random sample in the triangle.

> 5. How would you generate a random sample from the following statistical distribution?
>
>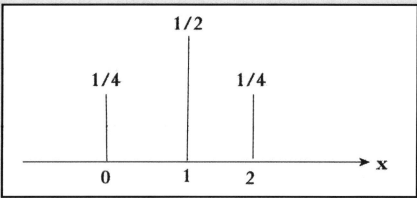
>
> Figure 42.3 Sampling from a binomial distribution

Solution: Generate a random number R between 0 and 1 and then compute

$$X = \begin{cases} 0 & 0 \leq R \leq 0.25 \\ 1 & 0.25 < R \leq 0.75 \\ 2 & 0.75 < R \leq 1 \end{cases}$$

The random variable X has the desired statistical (binomial) distribution.

> 6. The Buffon needle problem states that if a needle of a given length, say 1 inch, is tossed randomly on a flag (or any ruled surface) where the width between the stripes is the same as the length of the needle, i.e. 1 inch, then the probability the needle crosses a stripe is $2/\pi$. How would you devise a game to estimate the value of π?

Solution: This is a problem in what is called geometric probability. In this case we generate two random numbers r, θ where $0 \leq r \leq 1$ and $0 \leq \theta \leq \pi/2$. If

$$y_1 = r - \frac{1}{2}\sin\theta < 0$$

or

$$y_2 = r + \frac{1}{2}\sin\theta > 1$$

then the needle crosses a line, otherwise it does not. Perform this experiment n times and take the average $\hat{p}$ of the number of times the needle crosses a line, and then estimate π from the formula

$$\frac{2}{\pi} \doteq \hat{p} \implies \pi \doteq \frac{2}{\hat{p}}$$

Figure 42.4 illustrates the problem.

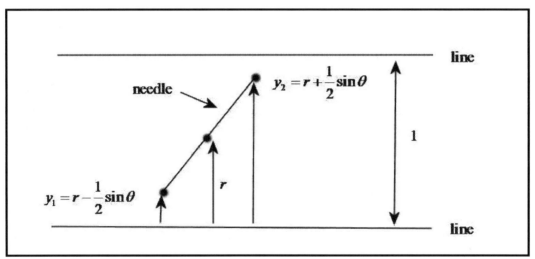

Figure 42.4 Diagram of the Buffon needle problem

Try writing your own program with a different number of tosses and see how close to π you come. In fact, make a graph of your approximation to π as a function of n, the number of tosses.

ΞΨΘΠΩΣ

Lesson 43: Monte Carlo Solution of Partial Differential Equations

> 1. Write a flow diagram for *tour du wino* that finds the approximate solution to the following BV problem at some interior grid point. Let the number of horizontal and vertical grid points be arbitrary.
>
> PDE: $u_{xx} + u_{yy} = 0$, $0 < x < 1$, $0 < y < 1$
>
> BC: $u(x,y) = g(x,y)$ on the boundary

Solution: Hint: One idea would be to set up an $n \times n$ grid matrix $G = (g_{i,j})$, $i, j = 1, 2, ..., n$ where increasing i is in the increasing y direction (when $i = 1$ we have $y = 0$ and when $i = n$ we have $y = 1$; also when $j = 1$ we have $x = 0$ and when $j = n$ we have $x = 1$. We also define another $n \times n$ array $P = (p_{i,j})$, $i, j = 1, 2, ..., n$, where initially all elements are set to zero, later the first and last rows, first and last columns will contain the fraction of times the random walks ends at the given boundary. Finally, we set up a matrix G where the first and last rows, and the first and last columns are the BCs $g(x, y)$. We leave the details of a program up to a good programmer.

> 2. Write a computer program to carry out the flow diagram from Problem 1
>
> PDE: $u_{xx} + x^2 u_{yy} = 0$, $0 < x < 1$, $0 < y < 1$
>
> BC: $u(x,y) = g(x,y)$ on the boundary

Solution: The finite-difference approximation to the derivatives in the PDE are

$$u_{xx} \doteq \frac{1}{h^2}\left(u_{i,j+1} - 2u_{i,j} + u_{i,j-1}\right)$$

$$u_{yy} \doteq \frac{1}{k^2}\left(u_{i+1,j} - 2u_{i,j} + u_{i-1,j}\right)$$

and so the PDE reduces to (letting $h = k$ for simplicity)

292 Lesson 43: Monte Carlo Solution of Partial Differential Equations

$$\frac{1}{h^2}\left(u_{i,j+1} - 2u_{i,j} + u_{i,j-1}\right) + \frac{x_j^2}{h^2}\left(u_{i+1,j} - 2u_{i,j} + u_{i-1,j}\right) = 0$$

and solving for $u_{i,j}$ gives

$$u_{i,j} = \frac{1}{2(x_j^2+1)}\left[u_{i,j+1} + u_{i,j-1} + x_j^2\left(u_{i+1,j} + u_{i-1,j}\right)\right]$$

Hence the probability of moving from $u_{i,j}$ to one of its four neighbors is the coefficient of the respective terms, which if you note are positive and

$$\frac{1}{2x_j^2+2} + \frac{1}{2x_j^2+2} + \frac{x_j^2}{2x_j^2+2} + \frac{x_j^2}{2x_j^2+2} = \frac{1}{2x_j^2+2}(2x_j^2+2) = 1$$

sum to one. We leave the problem of writing a computer program to the reader.

---

3. How would the game of tour du wino be modified to solve

$$\text{PDE: } u_{xx} + x^2 u_{yy} = 0,\ 0 < x < 1,\ 0 < y < 1$$

$$\text{BC: } u(x,y) = g(x,y) \quad \text{on the boundary}$$

Solution: Plugging the finite difference approximations

$$u_{xx} \doteq \frac{1}{h^2}\left(u_{i,j+1} - 2u_{i,j} + u_{i,j-1}\right)$$

$$u_{yy} \doteq \frac{1}{k^2}\left(u_{i+1,j} - 2u_{i,j} + u_{i-1,j}\right)$$

with $h = k$, into the PDE, we find (after brief algebra)

$$\frac{1}{h^2}\left(u_{i,j+1} - 2u_{i,j} + u_{i,j-1}\right) + \frac{x_j^2}{h^2}\left(u_{i+1,j} - 2u_{i,j} + u_{i-1,j}\right) = 0$$

and solving for $u_{i,j}$ gives

Lesson 43: Monte Carlo Solution of Partial Differential Equations

$$u_{i,j} = \frac{1}{2(x_j^2+1)}\left[u_{i,j+1} + u_{i,j-1} + x_j^2\left(u_{i+1,j} + u_{i-1,j}\right)\right]$$

Hence the probability of moving from $u_{i,j}$ to one of its four neighbors is the coefficient of the respective terms, which if you note are positive and

$$\frac{1}{2x_j^2+2} + \frac{1}{2x_j^2+2} + \frac{x_j^2}{2x_j^2+2} + \frac{x_j^2}{2x_j^2+2} = \frac{1}{2x_j^2+2}(2x_j^2+2) = 1$$

sum to one. We leave the problem of writing a computer program to the reader.

---

4. Devise a modified tour du wino game that will solve

    PDE: $u_{xx} + u_{yy} = 0$, $0 < x < 1$, $0 < y < 1$

    BCs: $u(x,y) = \begin{cases} u(x,1) = 0 \\ u(x,0) = 0 \\ u(0,y) = 1 \\ u_x(1,y) = 0 \end{cases}$ $0 < x < 1, 0 < y < 1$

Solution: We construct a grid of points for the unit square in Figure 43.1 which are shown in Figure 43.4. The BCs give rise to the following equations at the boundary grid points

$$u(0,y) = 1 \Rightarrow u_{i,1} = 1, \ i = 1,2,\ldots,n \quad \text{(left edge)}$$

$$u_x(1,y) = 0 \Rightarrow u_{i,n} - u_{i,n-1} = 0, \ i = 1,2,\ldots,n \quad \text{(right edge)}$$

$$u(x,0) = 0 \Rightarrow u_{1,j} = 0, \ j = 1,2,\ldots,n \quad \text{(bottom edge)}$$

$$u(x,1) = 0 \Rightarrow u_{n,j} = 0, \ j = 1,2,\ldots,n \quad \text{(top edge)}$$

294 Lesson 43: Monte Carlo Solution of Partial Differential Equations

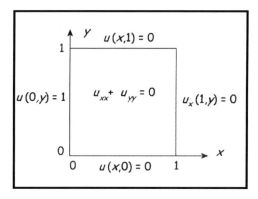

Figure 43.1

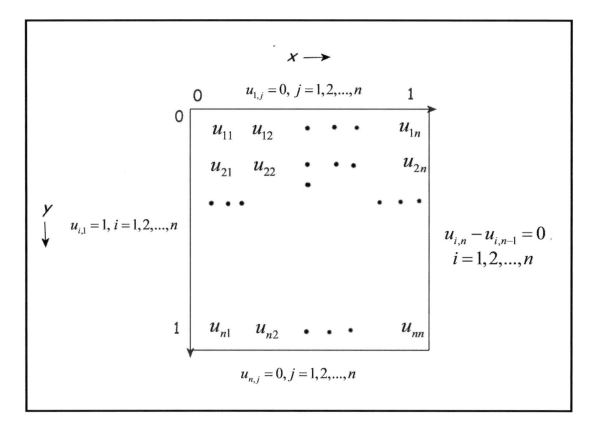

Figure 43.2 Grid of points for the unit square

Note that the derivative BC $u_x(1,y)=0$ becomes $u_{i,n}-u_{i,n-1}=0$ or $u_{i,n}=u_{i,n-1}$ which means that the wino arriving at the boundary grid point (i,n) has the same reward as it had the previous step at $(i,n-1)$. Hence, this gives us the strategy at boundaries with derivative BCs; when the wino hits such a boundary, like the right edge at $x=1$ in this problem, it simply bounces back to the interior. The wino may bounce back to the boundary,

Lesson 43: Monte Carlo Solution of Partial Differential Equations

but eventually it will move on to other interior points, eventually arriving at one with fixed BCs.

> 5. Derive the Monte Carlo game for solving the following parabolic IBVP.
>
> PDE: $u_t = u_{xx} - u_x - u \quad 0 < x < 1,\ 0 < y < 1$
>
> BC: $u(x,y) = \begin{cases} u(0,t) = f(t) \\ u(1,t) = g(t) \end{cases} \quad 0 < t < \infty$
>
> IC: $u(x,0) = \phi(x) \quad 0 \leq x \leq 1$

Solution: The Crank-Nicolson method gives us the finite difference equation

$$u_{i+1,j} = P_1 u_{i+1,j-1} + P_2 u_{i+1,j+1} + P_3 u_{i,j-1} + P_4 u_{i,j} + P_5 u_{i,j+1}$$

where the P_i^s are numbers computed from the coefficients in the PDE and represent the probabilities of moving to the corresponding point. That is

$$P_1 + P_2 + P_3 + P_4 + P_5 = 1$$

One can modify the BCs of this problem to solve problems with derivative BCs (i.e flux across the boundaries).

$$\Sigma Z \Psi \Xi \Omega \Sigma$$

Lesson 44: Calculus of Variations (Euler-Lagrange Equations)

> 1. Find the minimizing function $\bar{y}(x)$ of the functional
>
> $$J[y] = \int_0^1 \sqrt{1 + \left(\frac{dy}{dx}\right)^2}\, dx$$
>
> among all (differentiable) functions $y(x)$ satisfying the BC $y(0)=1$, $y(1)=1$. What is your interpretation of your answer? What is the value of $J[\bar{y}]$?

Solution: We are given the integrand

$$F(x, y, y') = \sqrt{1 + y'^2}$$

and so the Euler-Lagrange equation is

$$F_y - \frac{d}{dx} F_{y'} = -\frac{d}{dx}\left(\frac{y'}{\sqrt{1+y'^2}}\right) = 0$$

which implies

$$\frac{y'}{\sqrt{1+y'^2}} = c \Rightarrow y' = \frac{c}{\sqrt{1-c^2}} = A \Rightarrow y(x) = Ax + B$$

and applying the BCs $y(0) = 0$ and $y(1) = 1$, we have

$$\bar{y}(x) = x$$

Since $J[y]$ stands for the length of the curve $y(x)$ between two points, the answer is not surprising since it is a straight line. The value of the functional for this minimizing function is $J[x] = \sqrt{2}$ which is the length of the curve (line).

Lesson 44: Calculus of Variations (Euler-Lagrange Equations)

> 2. A vibrating mass attached to a spring has kinetic energy (KE) and potential energy (PE) of
>
> $$KE = \frac{1}{2}m\left(\frac{dy}{dt}\right)^2, \quad PE = \frac{1}{2}ky^2$$
>
> where
>
> $m =$ mass of the object
>
> $k =$ spring constant
>
> $y =$ position of the mass
>
> Hamilton's principle states that at all times the position $y(t)$ of the mass is such that the following integral
>
> $$\int_{t_1}^{t_2}[KE - PE]dt = \int_{t_1}^{t_2}\left[\frac{1}{2}m\left(\frac{dy}{dt}\right) - \frac{1}{2}ky^2\right]dt = \min$$
>
> is minimized. Show that if Hamilton's principle holds, then the position $y(t)$ of the mass must satisfy the differential equation
>
> $$m\frac{d^2y}{dt^2} + ky = 0$$

Solution: Differentiating the Lagrangian

$$L = KE - PE = \frac{1}{2}my'^2 - ky^2$$

with respect to y and y', we have

$$\frac{\partial L(x,y,y')}{\partial y} = \frac{\partial}{\partial y}\left[\frac{1}{2}my'^2 - \frac{1}{2}ky^2\right] = -ky$$

$$\frac{\partial L(x,y,y')}{\partial y'} = \frac{\partial}{\partial y'}\left[\frac{1}{2}my'^2 - \frac{1}{2}ky^2\right] = my'$$

Plugging these values into the Euler-Lagrange equations, we get

Lesson 44: Calculus of Variations (Euler-Lagrange Equations)

$$L_y - \frac{d}{dx}L_{y'} = -ky - \frac{d}{dx}(my') = -ky - my'' = 0$$

Hence, it is not surprising that we have the undamped harmonic oscillator $my'' + ky = 0$.

---

3. Show that the minimizing function $\bar{y}(x)$ of the functional

$$J[y] = \int_0^{\pi/2} \left[y'^2 - y^2 \right] dx$$

among all (smooth) functions satisfying the BC $y(0) = 0$, $y(\pi/2) = 1$ is $\bar{y}(x) = \sin x$ and evaluate $J[\sin x]$.

Solution: We differentiate

$$L_y = \frac{\partial}{\partial y}\left[y'^2 - y^2 \right] = -2y$$

$$L_{y'} = \frac{\partial}{\partial y'}\left[y'^2 - y^2 \right] = 2y'$$

Plugging these in the Euler-Lagrange equation, gives

$$L_y - \frac{d}{dx}L_{y'} = -2y - \frac{d}{dx}(2y') = -2y - 2y'' = 0$$

or

$$y'' + y = 0$$

which has the general solution

$$y(x) = c_1 \cos x + c_2 \sin x$$

Plugging this in the BCs $y(0) = 0$, $y(\pi/2) = 1$ gives $y(x) = \sin x$. The value of the functional is

Lesson 44: Calculus of Variations (Euler-Lagrange Equations)

$$J[y] = \int_0^{\pi/2} \sin x \, dx = -\cos x \Big|_0^{\pi/2} = 1$$

> **4.** Derive the Euler-Lagrange equation
>
> $$F_u - \frac{\partial}{\partial x} F_{u_x} - \frac{\partial}{\partial y} F_{u_y} = 0$$
>
> for the functional
>
> $$J[u] = \iint_D F(x, y, u, u_x, u_y) \, dx \, dy$$

Solution: The strategy of finding the minimizing function $\bar{u}(x, y)$ is to take the derivative of

$$\phi(\varepsilon) = J\big[\bar{u}(x, y) + \varepsilon \eta(x, y)\big]$$

with respect to ε and set the derivative to zero. That is

$$\begin{aligned}
\frac{d\phi(\varepsilon)}{d\varepsilon} &= \frac{d}{d\varepsilon} J[\bar{u} + \varepsilon \eta]\Big|_{\varepsilon=0} \\
&= \frac{d}{d\varepsilon} \iint_D F(x, y, \bar{u} + \varepsilon \eta, \bar{u}_x + \varepsilon \eta_x, \bar{u}_y + \varepsilon \eta_y) \, dx \, dy \\
&= \iint_D \left(\frac{\partial F}{\partial \bar{u}} \eta + \frac{\partial F}{\partial \bar{u}_x} \eta_x + \frac{\partial F}{\partial \bar{u}_y} \eta_y \right) dx \, dy \\
&= \iint_D \left[\frac{\partial F}{\partial \bar{u}} - \frac{\partial}{\partial x}\left(\frac{\partial F}{\partial \bar{u}_x}\right) - \frac{\partial}{\partial y}\left(\frac{\partial F}{\partial \bar{u}_y}\right) \right] \eta(x, y) \, dx \, dy \\
&= 0
\end{aligned}$$

where the last step in the above equations is simply integration by parts, i.e. taking the derivatives off η_x and η_y and putting them on

$$\frac{\partial F}{\partial \bar{u}_x} \quad \text{and} \quad \frac{\partial F}{\partial \bar{u}_y}$$

respectively. But in order that the above integral is zero for an arbitrary function $\eta(x,y)$, we have the other factor in the integrand to be zero. That is

$$\frac{\partial F}{\partial \overline{u}} - \frac{\partial}{\partial x}\left(\frac{\partial F}{\partial \overline{u}_x}\right) - \frac{\partial}{\partial y}\left(\frac{\partial F}{\partial \overline{u}_y}\right) = 0$$

ΞΨΩΘΠΣ

Lesson 45: Variational Methods for Solving PDEs (Method of Ritz)

> 1. What would be the energy functional $J[u]$ corresponding to the following problem.
>
> PDE: $u_{xx} + u_{yy} = 1 \quad 0<x<1,\ 0<y<1$
>
> BC: $u(x,y)=0$ on the boundary

Solution: In Lesson 44 we saw that the Euler-Lagrange equation for the functional

$$J[u] = \iint_D F(x,y,u,u_x,u_y)\,dx\,dy$$

was

$$F_u - \frac{\partial}{\partial x}F_{u_x} - \frac{\partial}{\partial y}F_{u_y} = 0$$

The goal in this problem is to go backwards, i.e. find the functional whose Euler-Lagrange equation is the given PDE

$$u_{xx} + u_{yy} = 1$$

There really isn't a simple, procedure for finding the energy functional of a PDE, but by trial and error

$$J[u] = \int_0^1 \int_0^1 \left[u_x^2 + u_y^2 + 2u \right] dx\,dy$$

we get

$$F = u_x^2 + u_y^2 + u$$
$$F_u = 2$$
$$F_{u_x} = 2u_x$$
$$F_{u_y} = 2u_y$$

which gives the Euler-Lagrange equation

Lesson 45: Variational Methods for Solving PDEs (Method of Ritz)

$$F_u - \frac{\partial}{\partial x}F_{u_x} - \frac{\partial}{\partial y}F_{u_y} = 2 - \frac{\partial}{\partial x}(2u_x) - \frac{\partial}{\partial y}(2u_y)$$
$$= 2 - 2(u_{xx} + u_{yy})$$
$$= 0$$

or the given Poisson equation

$$u_{xx} + u_{yy} = 1$$

---

2. How could we minimize the following functional by the method of Ritz?

$$J[y] = \int_0^1 \left[y^2(x) + \left(\frac{dy}{dx}\right)^2 \right] dx \quad y(0)=0,\ y(1)=1$$

Hint: Since the function $y(t)$ does not have zero BC we introduce the new function $z(x)$ defined by

$$z(x) = (1-x)y(x)$$

which converts the BCs to $z(0) = z(1) = 0$

Solution: We could let

$$z(x) = (1-x)y(x)$$

which gives

$$z'(x) = -y(x) + (1-x)y'(x)$$

and solve for y and y', then plug these values in the original functional $J[y]$ getting

$$J[z] = \int_0^1 \left[\left(\frac{z}{1-x}\right)^2 + \left(\frac{z'}{1-x} + \frac{z}{(1-x)^2}\right)^2 \right] dx$$

Since $z(0) = z(1) = 0$. After we find $\bar{z}$ we find $\bar{y}(x)$ from

Lesson 45: Variational Methods for Solving PDEs (Method of Ritz)

$$\bar{y} = \frac{\bar{z}}{1-x}$$

The point is we can use the Method of Ritz since it satisfies the BC $z(0) = z(1) = 0'$

We could also replace the function $y(x)$ in the functional $J[y]$ with the approximation

$$y_n(x) = a_1 x + a_2 x^2 + \cdots + a_n x^n$$

which results in the functional taking the form

$$\phi(a_1, a_2, \ldots, a_n) = \int_0^1 \left[\left(\sum_{k=1}^n a_k x^k \right)^2 + 2 \left(\sum_{k=1}^n k a_k x^{k-1} \right)^2 \right] dx$$

The minimum could then be found by solving for the coefficients $a_k, k = 1, 2, \ldots, n$ from the equations

$$\frac{\partial \phi}{\partial a_k} = 0, \ k = 1, 2, \ldots, n$$

---

3. Write a computer program to carry out the computations in Figure 25.2.

Solution: Student project.

---

4. Show that the Euler-Lagrange equation for the functional

$$J[u] = \int_0^1 \int_0^1 \left[u_x^2 + u_y^2 \right] dx dy$$

is

$$u_{xx} + u_{yy} = 0$$

Solution: The proof is similar to the proof in Lesson 44 when we found the Euler-Lagrange equation for a function $y(x)$ of one variable. In this case we consider the function of ε defined by

Lesson 45: Variational Methods for Solving PDEs (Method of Ritz)

$$\phi(\varepsilon) = J[\bar{u} + \varepsilon\eta]$$

where $\bar{u}(x,y)$ is the minimizing function; i.e. the function that minimizes the functional $J[u]$, and $\eta(x,y)$ is a small arbitrary (differentiable) variation around $\bar{u}$ which is zero on the boundaries of the square. The minimizing function $\bar{u}$ can be found by setting the derivative (with respect to ε) of $\phi(\varepsilon)$ to zero at $\varepsilon = 0$ and then solving for $\bar{u}$. Doing this, gives

$$\begin{aligned}
\frac{d\phi(\varepsilon)}{d\varepsilon} &= \frac{d}{d\varepsilon} J[\bar{u} + \varepsilon\eta]\bigg|_{\varepsilon=0} \\
&= \frac{d}{d\varepsilon} \int_0^1 \int_0^1 \left[F(x, y, \bar{u} + \varepsilon\eta, \bar{u}_x + \varepsilon\eta_x, \bar{u}_y + \varepsilon\eta_y) \right] dxdy \\
&= \int_0^1 \int_0^1 \left[\frac{\partial F}{\partial \bar{u}} \eta(x) + \frac{\partial F}{\partial \bar{u}_x} \eta_x(x) + \frac{\partial F}{\partial \bar{u}_y} \eta_y(x) \right] dxdy \\
&= \int_0^1 \int_0^1 \left[\frac{\partial F}{\partial \bar{u}} - \frac{d}{dx}\left(\frac{\partial F}{\partial \bar{u}_x}\right) - \frac{d}{dy}\left(\frac{\partial F}{\partial \bar{u}_y}\right) \right] \eta(x,y) dxdy = 0
\end{aligned}$$

Now, if the above integral is zero for an arbitrary function $\eta(x,y)$, the function $\bar{u}$ must satisfy

$$\frac{\partial F}{\partial \bar{u}} - \frac{d}{dx}\left(\frac{\partial F}{\partial \bar{u}_x}\right) - \frac{d}{dy}\left(\frac{\partial F}{\partial \bar{u}_y}\right) = 0$$

Hence, if $\bar{u}(x,y)$ minimizes the functional

$$J[u] = \int_0^1 \int_0^1 F(x, y, u, u_x, u_y) \, dxdy$$

then $\bar{u}(x,y)$ must satisfy the PDE (Euler-Lagrange equation)

$$\frac{\partial F}{\partial u} - \frac{d}{dx}\left(\frac{\partial F}{\partial u_x}\right) - \frac{d}{dy}\left(\frac{\partial F}{\partial u_y}\right) = 0$$

Lesson 45: Variational Methods for Solving PDEs (Method of Ritz)

In the current problem the functional is

$$J[u] = \int_0^1 \int_0^1 \left[u_x^2 + u_y^2\right] dx dy$$

and so

$$F(x, y, u, u_x, u_y) = u_x^2 + u_y^2$$

hence the Euler-Lagrange equation is

$$\frac{\partial F}{\partial u} - \frac{d}{dx}\left(\frac{\partial F}{\partial u_x}\right) - \frac{d}{dy}\left(\frac{\partial F}{\partial u_y}\right) = -\frac{d}{dx}(2u_x) - \frac{d}{dy}(2u_y)$$

$$= -2u_{xx} - 2u_{yy} = 0$$

or Laplace's equation

$$u_{xx} + u_{yy} = 0$$

---

5. The following Dirichlet problem can be solved

PDE: $u_{xx} + u_{yy} = \sin(\pi x)$ $0 < x < 1,\ 0 < y < 1$

BC: $u(x, y) = 0$ on the boundary of the square

can be solved by the finite sine transform (transforming the x variable) and has the solution

$$u(x, y) = \left[0.06 e^{\pi y} + 0.04 e^{-\pi y} - \frac{1}{\pi^2}\right] \sin(\pi x)$$

How would you find the potential energy of this system?

Solution: The finite sine transform is

$$\text{ODE} \quad \frac{d^2 U_n(y)}{dy^2} - (n\pi)^2 U_n = \begin{cases} 1 & n = 1 \\ 0 & n = 2, 3, \ldots \end{cases}$$

$$\text{BCs} \quad \begin{cases} U_n(0) = 0 \\ U_n(1) = 0 \end{cases}$$

Lesson 45: Variational Methods for Solving PDEs (Method of Ritz)

which has the solution

$$U_1(y) = Ae^{\pi y} + Be^{-\pi y} - \frac{1}{\pi^2}$$

$$U_n(y) = 0, \quad n = 2, 3, \ldots$$

We can now take the inverse sine transform of these functions to get $u(x, y)$ after which the potential energy is

$$J[u] = \int_0^1 \int_0^1 \left[u_x^2 + u_y^2 + 2u \sin(\pi x) \right] dx \, dy$$

Most likely this integral would have to be evaluated numerically.

$$\Psi\Omega\Xi\Theta\Pi\Sigma$$

Lesson 46: Perturbation Methods for Solving PDEs

> 1. Substitute
> $$u = u_0 + \varepsilon u_1 + \varepsilon^2 u_2$$
> into the BVP
> $$\text{PDE: } u_{rr} + \frac{1}{r} u_r + \frac{1}{r^2} u_{\theta\theta} = 0 \quad 0 < r < 1 + \varepsilon \sin\theta$$
> $$\text{BC: } u(1,\theta) + u_r(1,\theta)(\varepsilon \sin\theta) + u_{rr}(1,\theta)\left(\frac{\varepsilon \sin\theta}{2!}\right)^2 + \cdots = \cos\theta$$
> to find a sequence of problems $P_0, P_1, P_2, \ldots$.

Solution: We use only the first two terms $u = u_0 + \varepsilon u_1$ in the perturbation series since the calculations become long and tedious, although the ideas are the same. Substituting this sum into the PDE and first two terms of the BC expansion, gives

$$\text{PDE: } (u_0 + \varepsilon u_1)_{rr} + \frac{1}{r}(u_0 + \varepsilon u_1)_r + \frac{1}{r^2}(u_0 + \varepsilon u_1)_{\theta\theta} = 0$$

$$\text{BC: } (u_0(1,\theta) + \varepsilon u_1(1,\theta)) + \left(\frac{\partial u_0(1,\theta)}{\partial r} + \varepsilon \frac{\partial u_1(1,\theta)}{\partial r}\right)(\varepsilon \sin\theta) = \cos\theta$$

Now, setting the coefficients of 1 and ε equal to each other on each side of the equations, we find the problems P_0 and P_1, respectively

$$P_0 \begin{cases} \text{PDE: } \dfrac{\partial^2 u_0}{\partial r^2} + \dfrac{1}{r}\dfrac{\partial u_0}{\partial r} + \dfrac{1}{r^2}\dfrac{\partial^2 u_0}{\partial \theta^2} = 0 \\ \text{BC: } u_0(1,\theta) = \cos\theta \end{cases}$$

$$P_1 \begin{cases} \text{PDE: } \dfrac{\partial^2 u_1}{\partial r^2} + \dfrac{1}{r}\dfrac{\partial u_1}{\partial r} + \dfrac{1}{r^2}\dfrac{\partial^2 u_1}{\partial \theta^2} = 0 \\ \text{BC: } u_1(1,\theta) = -\dfrac{\partial u_0(1,\theta)}{\partial r}\sin\theta \end{cases}$$

Note that the solution to P_0, which is $u_0(r,\theta) = r\cos\theta$ is used in problem P_1. Hence problem P_1 becomes

$$P_1 \begin{cases} \text{PDE: } \dfrac{\partial^2 u_1}{\partial r^2} + \dfrac{1}{r}\dfrac{\partial u_1}{\partial r} + \dfrac{1}{r^2}\dfrac{\partial^2 u_1}{\partial \theta^2} = 0 \\ \text{BC: } u_1(1,\theta) = -\cos\theta \sin\theta = -\dfrac{1}{2}\sin(2\theta) \end{cases}$$

which has the solution

$$u_1(r,\theta) = -\frac{1}{2}r^2 \sin(2\theta)$$

Hence, the first two terms in the perturbation expansion $(\varepsilon = 1)$ are

$$u(r,\theta) = u_0(r,\theta) + u_1(r,\theta) = r\cos\theta - \frac{1}{2}r^2 \sin(2\theta)$$

---

2. Show that the following nonlinear problem

$$\text{PDE: } u_{rr} + \frac{1}{r}u_r + \frac{1}{r^2}u_{\theta\theta} + u^2 = 0 \quad 0 < r < 1$$

$$\text{BC: } u(1,\theta) = \cos\theta \quad 0 \leq \theta < 2\pi$$

discussed in the text gives rise to the sequence of linear problems

$$P_0 \begin{cases} \nabla^2 u_0 = 0 & 0 < r < 1 \\ u_0(1,\theta) = \cos\theta \end{cases} \Rightarrow u_0(r,\theta) = r\cos\theta$$

$$P_1 \begin{cases} \nabla^2 u_1 = -u_0^2 & 0 < r < 1 \\ u_1(1,\theta) = 0 \end{cases}$$

$$P_2 \begin{cases} \nabla^2 u_2 = -2u_0 u_1 & 0 < r < 1 \\ u_2(1,\theta) = 0 \end{cases}$$

$$\cdots \quad \cdots$$

Solution: We seek a solution of the form

Lesson 46: Perturbation Methods for Solving PDEs

$$u = u_0 + \varepsilon u_1 + \varepsilon^2 u_2$$

and plugging this into the PDE and BC gives

$$\left(u_0 + \varepsilon u_1 + \varepsilon^2 u_2\right)_{rr} + \frac{1}{r}\left(u_0 + \varepsilon u_1 + \varepsilon^2 u_2\right)_r + \frac{1}{r^2}\left(u_0 + \varepsilon u_1 + \varepsilon^2 u_2\right)_{\theta\theta} + \varepsilon\left(u_0 + \varepsilon u_1 + \varepsilon^2 u_2\right)^2 = 0$$

$$u_0(1,\theta) + \varepsilon u_1(1,\theta) + \varepsilon^2 u_2(1,\theta) = \cos\theta$$

Now, setting the coefficients of $1, \varepsilon$ and ε^2 equal to each other on each side of the equations, we find the problems P_0, P_1 and P_2, respectively

$$P_0 \begin{cases} \text{PDE: } \dfrac{\partial^2 u_0}{\partial r^2} + \dfrac{1}{r}\dfrac{\partial u_0}{\partial r} + \dfrac{1}{r^2}\dfrac{\partial^2 u_0}{\partial \theta^2} = 0 \\ \text{BC: } u_0(1,\theta) = \cos\theta \end{cases} \Rightarrow u_0(r,\theta) = r\cos\theta$$

$$P_1 \begin{cases} \text{PDE: } \dfrac{\partial^2 u_1}{\partial r^2} + \dfrac{1}{r}\dfrac{\partial u_1}{\partial r} + \dfrac{1}{r^2}\dfrac{\partial^2 u_1}{\partial \theta^2} + u_0^2 = 0 \\ \text{BC: } u_1(1,\theta) = 0 \end{cases}$$

$$P_2 \begin{cases} \text{PDE: } \dfrac{\partial^2 u_1}{\partial r^2} + \dfrac{1}{r}\dfrac{\partial u_1}{\partial r} + \dfrac{1}{r^2}\dfrac{\partial^2 u_1}{\partial \theta^2} + 2u_0 u_1 = 0 \\ \text{BC: } u_1(1,\theta) = 0 \end{cases}$$

Although problems P_1 and P_2 are linear they are not homogeneous and not trivial. However, they can be solved and when done, we will have a good analytical approximation

$$u = u_0 + u_1 + u_2$$

---

3. To check for accuracy in our perturbation approximation, substitute

$$u(r,\theta) = u_0 + u_1$$

$$= r\cos\theta - \frac{(r^4 - 1)}{32} - \frac{(r^4 - r^2)}{24}\cos(2\theta)$$

---

in the nonlinear problem

$$\text{PDE: } \nabla^2 u + u^2 = 0 \quad 0 < r < 1$$
$$\text{BC: } u(1,\theta) = \cos\theta \quad 0 \leq \theta < 2\pi$$

Solution: If we substitute

$$u_0 + u_1 = r\cos\theta + \left(\frac{r^4 - 1}{32}\right) - \left(\frac{r^4 - r^2}{24}\right)\cos(2\theta)$$

into $\nabla^2 u + u^2$ and $u(1,\theta)$ we will see that $u(1,\theta) = 0$ and that the function $\nabla^2 u + u^2$ is very small inside the unit circle $0 < r < 1$. The term r^2 will contain 9 terms but can be evaluated nevertheless.

4. Solve the problem

$$P_1 \begin{cases} \nabla^2 u_1 = 0 \quad 0 < r < 1 \\ u_1(1,\theta) = -\sin\theta\left(\dfrac{\partial u_0}{\partial r}\right) = -\sin\theta\cos\theta \end{cases}$$

in the boundary pertubation problem discussed in the book to see how well the approximation

$$u(r,\theta) = u_0(r,\theta) + \frac{1}{4}u_1(r,\theta)$$

satisfies

$$\text{PDE: } \nabla^2 u = 0 \quad 0 < r < 1$$
$$\text{BC: } u\left(1 + \frac{1}{4}\sin\theta, \theta\right) = \cos\theta \quad 0 \leq \theta < 2\pi$$

Solution: We begin by making the transformation

$$u(r,\theta) = \sin\theta\cos\theta + U(r,\theta)$$

to a new function $U(x,\theta)$ which will satisfy the new problem

Lesson 46: Perturbation Methods for Solving PDEs

$$\text{PDE: } \nabla^2 U = \frac{1}{r^2} \quad 0 < r < 1$$

$$\text{BC: } U(1,\theta) = 0 \quad 0 \le \theta < 2\pi$$

We now solve this problem by letting

$$U(r,\theta) = U_h(r,\theta) + U_p(r,\theta)$$

where $U_h(r,\theta)$ is the general solution of the homogeneous equation $\nabla^2 U = 0$ and $U_p(r,\theta)$ is any single solution (called a particular solution) of the nonhomogeneous equation $\nabla^2 U = 1/r^2$. A general form of the solution of the homogeneous equation $\nabla^2 U = 0$ (Laplace's equation) is

$$U_h(r,\theta) = \sum_{n=1}^{\infty} \left[a_n \cos(n\theta) + b_n \sin(n\theta) \right]$$

To find a particular solution, we try a solution of the form

$$U_p(r,\theta) = A \cos(2\theta)$$

and find $A = -1/4$. Hence, we have

$$U(r,\theta) = U_h(r,\theta) + U_p(r,\theta)$$
$$= \sum_{n=1}^{\infty} \left[a_n \cos(n\theta) + b_n \sin(n\theta) \right] - \frac{1}{4} \cos(2\theta)$$

If we now plug this value into the BC $U(1,\theta) = 0$ we get $a_2 = 1/4$ all the other a_n^s, b_n^s are zero. Hence, we have the solution

$$U(r,\theta) = \frac{1}{4} r^2 \cos(2\theta) - \frac{1}{4} \cos(2\theta)$$

or

$$u(r,\theta) = \sin\theta \cos\theta + \frac{1}{4} r^2 \cos(2\theta) - \frac{1}{4} \cos(2\theta)$$

One would now have to use a computer to evaluate this function to see how well it satisfies the BC

$$\text{BC: } u\left(1+\frac{1}{4}\sin\theta, \theta\right) = \cos\theta \quad 0 \leq \theta < 2\pi$$

ΞΨΖΥΘΩ

Lesson 47: Conformal Mapping Solutions of PDEs

> 1. Where is the following mapping conformal?
>
> $$w = u + iv = \log\left[\frac{z-1}{z+1}\right]$$
>
> Convince yourself that the mapping maps the upper half z plane into the strip $-\infty < u < \infty, 0 < v < \pi$ in the w plane.

Solution: We have

$$w = \log\left[\frac{z-1}{z+1}\right] \Rightarrow \frac{dw}{dz} = \frac{2}{z^2 - 1}$$

and so the mapping is conformal except at $z = \pm 1$. Now, from the fact that

$$w = \log z = \log\left(re^{i\theta}\right)$$
$$= \log r + \log e^{i\theta}$$
$$= \log r + i\theta \log e$$
$$= \log r + i\theta$$
$$= \log|z| + i\arg(z)$$

hence, we have

$$w = \log\left[\frac{z-1}{z+1}\right] = \log\left|\frac{z-1}{z+1}\right| + i\arg\left[\frac{z-1}{z+1}\right]$$

We see that the real part

$$u = \log\left|\frac{z-1}{z+1}\right|$$

can get as close to $-\infty$ as we wish by letting $z \to 1$ and as close to $+\infty$ as we wish by letting $z \to -1$. Hence, $-\infty < u < \infty$. On the other hand, the complex part

Lesson 47: Conformal Mapping Solutions of PDEs

$$w = \arg\left[\frac{z-1}{z+1}\right] = \arg(z-1) - \arg(z+1)$$

Now, if you look carefully at the drawing in Figure 47.1 you will see that for a complex number in the upper half complex plane that the argument (or angle) of $z-1$ (angle between $z=1$ and the positive x axis) will be equal or larger than the argument of $z+1$ (angle between $z=-1$ and the positive x axis). If you look carefully, the maximum this difference will be is when z lies on the positive complex axis and gets close to the origin we see that $\arg(z-1) \to \pi$ and $\arg(z+1) \to 0$ and so

$$\arg(z-1) - \arg(z+1) \to \pi$$

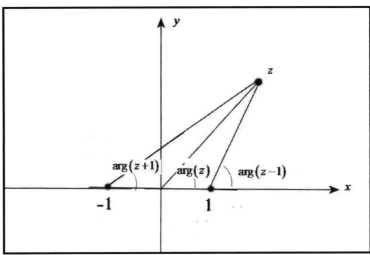

Figure 47.1 Graph showing angles of $z+1$ and $z-1$

---

2. What is the image of the first quadrant under the mapping $w = z^2$? It might be helpful to write the complex numbers in polar form

$$z = re^{i\theta}$$

Solution: Write

$$w = z^3 = \left[re^{i\theta}\right]^3 = r^3 e^{3i\theta}$$

and so the first quadrant maps into quadrants 1, 2, and 3.

Lesson 47: Conformal Mapping Solutions of PDEs

3. Solve the following Dirichlet problem in the first quadrant by means of the conformal mapping $w = z^2$.

PDE: $\phi_{xx} + \phi_{yy} = 0 \quad 0 < x < \infty, \; 0 < y < \infty$

BCs:
$$\begin{cases} \phi(x,0) = \begin{cases} 1 & 0 < x < 1 \\ 0 & 1 < x < \infty \end{cases} \\ \phi(0,y) = \begin{cases} 1 & 0 < y < 1 \\ 0 & 1 < y < \infty \end{cases} \end{cases}$$

Solution: Applying the transformation

$$w = z^2 = (x+iy)^2 = (x^2 - y^1) + i(2xy)$$

or in terms of two real transformations

$$u = x^2 - y^2$$
$$v = 2xy$$

The first quadrant is mapped into the upper half plane and the BCs mapped onto the real axis

PDE: $\phi_{uu} + \phi_{vv} = 0 \quad -\infty < u < \infty, \; 0 < v < \infty$

BCs:
$$\begin{cases} \phi(u,0) = 0 & -\infty < u < -1 \\ \phi(u,0) = 1 & -1 \le u \le 1 \\ \phi(u,0) = 0 & 1 < u < \infty \end{cases}$$

This problem has the solution

$$u(u,v) = \frac{1}{\pi} \tan^{-1}\left(\frac{2v}{u^2 + v^2 - 1} \right)$$

or in terms of x, y

$$u(x,y) = \frac{1}{\pi}\tan^{-1}\left[\frac{4xy}{\left(x^2-y^2\right)^2 + 4x^2y^2 - 1}\right]$$

4. Solve the following mixed Neumann problem inside a 45° wedge.

PDE: $\phi_{rr} + \frac{1}{r}\phi_r + \frac{1}{r^2}u_{\theta\theta} = 0 \quad 0 < r < 1,\ 0 \le \theta, 2\pi$

BCs: $\begin{cases} \phi(r,0) = 0 \\ \phi(r,\pi/4) = 1 \\ \phi_r(1,\theta) = 0 \end{cases}$

Hint: The complex function

$$f(z) = \log z = \log|z| + i\arg(z)$$

maps the ray $\theta = c_1$ in the z complex plane to the line $v = c_1$ in the w complex plane. It also maps the circle $r = c_2$ in the z plane to the straight line $u = \log c_2$ in the w plane, where log means the natural logarithm. The diagrams in the book might be helpful for visualizing this problem.

Solution: If we define the transformation that maps $z = x + iy$ to $w = u + iv$ by

$$w = \log z = \log|z| + i\arg(z)$$

or equivalently as the two real transformations

$$u = \log|z| = \log r$$
$$v = \arg(z) = \tan^{-1}(y/x)$$

we have that the transformation maps the 45 degree wedge in the problem to the region drawn in Figure 47.2.

Lesson 47: Conformal Mapping Solutions of PDEs

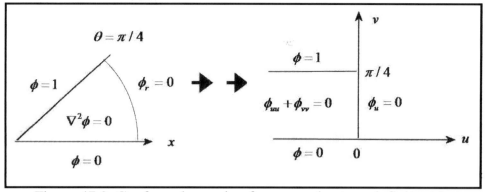

Figure 47.2 Conformal mapping from a wedge to an infinite strip

It is clear that the transformed problem has the solution

$$\phi(u,v) = \frac{4}{\pi}v$$

or in terms of the original coordinates x, y we have

$$\phi(x,y) = \frac{4}{\pi}\tan^{-1}\left(\frac{y}{x}\right)$$

ΞΨZΥΠΘ

How I (Almost) Proved Fermat's Last Theorem

At the time I thought it rather strange. After all, any alien space jockey worth his plasma knows he's supposed to touch down in an alligator swamp in South Georgia and scare the holy be-jesus out of some Georgia hillbillies. This time, however, they decided to land in the Maine woods and scare the holy be-jesus out of us Maine *woodsbillies*.

I was taking my evening stroll behind my house when suddenly I saw it, the unearthly green iridescent glow through the trees, the oval-shaped saucer hovering above the clearing. I hid behind some bushes and watched as a small opening materialized on the underbelly of the capsule and several otherworldly creatures, each no more than three-feet tall, emerged. Each had a human-like frame but with unmistakably alien features, a rubberlike skin that stretched over a bony scaffolding and two huge eyes that radiated an eerie green glow from a pasty-white face. I watched as no less than a dozen of these creatures made their way down a ladder. They all appeared similar except for one, which had a row of medals on its chest. I took it to be their leader.

After they reached the ground, the leader looked in my direction and shouted, "Farlow, we're here to make a deal." My body froze. "What's wrong, Farlow, is this the first time you've ever seen *Orkians* before?" the leader said as the creatures surrounded me. One of the creatures waved his bony little fingers before my eyes and suddenly my fear was gone. "That's better," the leader said. "Now let's get down to business."

"Business?" I stammered.

"You're a mathematician, aren't you Farlow?" the creature said. "And by the way, we don't like to be thought of as creatures, you know we can read minds, we prefer aliens. My name's Harry."

"Yes, Harry," I said. "I'm a mathematician."

"And not a very good one, we understand. This makes you the perfect candidate," the little creature said. A sharp pain ran through my body.

"I warned you, Farlow. We have ways for dealing with thoughts like those."

"I won't let it happen again, Harry," I moaned, "but I'm an excellent mathematician."

"Don't give us that," Harry said. "We know you haven't made a single discovery in your entire mathematical career."

"What about my proof of the *Widdlestein Conjecture?*" I protested.

"You call that a discovery?" Harry laughed. "Our kids learn that in kindergarten. But don't worry Farlow, we're going to give you the answer to the most famous mathematical problem in the world. We're going to make you a star."

"Wow!" I started to like the little guys.

"Just a few gallons of plasma should do," Harry said.

"What?" I asked.

"One discovery for just a little plasma," Harry said, "You didn't think we'd give you the answer to the most famous mathematical problem in the world for nothing, did you? A couple of gallons should do it right guys?" The other aliens nodded in unison.

"Wait," I protested. But before I could say anything the alien with the bony fingers waved them before my eyes.

The next thing I knew I was stretched out on a metallic table looking up at a small alien in a white lab coat. He was shining a beam of light in my face with a strange unearthly object which, for lack of a better description, looked like an ordinary flashlight. I panicked and screamed out.

"What's wrong, Farlow?" Harry asked. "Haven't you ever seen a flashlight before. Stop playing around with that damn thing, Raymond. And take that doctor's coat off. Damn kids," Harry said.

My body lie enmeshed in a tangle of wires and tubes. Ooze of several degrees of *yeeeeuuuuck* flowed like sea water through the tubes. I noticed that one of the tubes seemed to end somewhere in an opening in my abdomen. "What's all the gunk in that tube?" I asked.

"Oh don't mind that," Harry said. "It's just something we add."

"What about our deal?" I protested, trying to get up. "You said you were going to make me a famous mathematician."

"Yeah, yeah," Harry said. He then crossed the room to a filing cabinet and removed a large manila envelope from the top drawer. "Here it is," Harry said. "The answer to your most famous mathematical problem, a problem you earth people have been trying to solve for 358 years, a proof of Fermat's Last Theorem."

"What?!" I yelled, struggling to get up. "I'm giving you my plasma for Fermat's Last Theorem? That problem was solved 20 years ago!"

"What?" Harry seemed surprised.

"That's right," I said. "An Englishman proved it."

"Oh, sorry about that," Harry said. "We've been on the road so long you know. When we left they told us you didn't have a clue. What is that old theorem anyway?" Harry asked.

"I thought you were so smart," I said sarcastically.

"It's been a long time since kindergarten," Harry said.

"Well, let's start with something on that level then," I said. I thought it best if I gave Harry a beginner's lesson on the problem.

"You'll agree that

$$3^2 + 4^2 = 5^2$$

don't you?" I asked.

"Our kids ... ,"

"Yeah, yeah," I interrupted. "They learn it in kindergarten"

I also told Harry there are other pairs of numbers whose sum of squares is the square of a third number, such as

$$5^2 + 12^2 = 13^2$$

and

$$8^2 + 15^2 = 17^2$$

"What does all this have to do with Fermat's Last Theorem?" Harry asked, starting to get bored. I told Harry that although there are other integers like the above that satisfy the equation, the French mathematician, Pierre de Fermat, claimed there weren't any non-zero ones that satisfied

$$a^n + b^n = c^n$$

when the exponent *n* is greater than 2, like 3, 4, ... Fermat scribbled this claim in the margin of a book but said his proof would not fit in the margin. And so for 358 years many of the world's greatest mathematicians tried without success to prove Fermat's claim, known as Fermat's Last Theorem.

Finally, in 1994 a 40-year old English mathematician from Princeton University, Andrew Wiles, solved the problem.

"If you would have come 20 years ago, I'd been famous," I yelled at Harry. "*Harry, where the devil are you?*"

I then realized the aliens weren't the slightest interested in Fermat's Last Theorem, and were all clustered around a huge apparatus. Suddenly, I heard a voice cry out, "*How do you like the new me, Farlow?*" I looked up and saw the most hideous looking alien I'd ever seen, it was horrible. *I then realized I was looking in a mirror!*

"*Aaaaaaaagggggggghhhhhhhhh,*" I screamed. WHAT HAVE YOU DONE TO ME?!" Looking around the room, I saw a dozen spitting images Brad Pitt.

"It's amazing what a little plasma will do for your complexion," one of the Brad Pitts said in a voice that sounded an awfully lot like Harry.

"Everyone wants to be a Brad Pitt, you'd think someone would pick a Clooney or a Redford."

"*Aaaaaagggggggggggghhhhhhhhh,*" I screamed out again. "No, no, ... give me back my face, keep the damn theorem ... ."

".... wake up dear, you're dreaming again," someone yelled at me. I found myself sitting in my own bed, drenched in sweat. My wife was shaking me.

I looked at her but questions remained. Had I been dreaming, or had I actually been visited by creatures, OUCH, aliens? If so, might they return and give me the answer to the Riemann Hypothesis or what about the Goldbach

Conjecture? Only time would tell. I could still be a famous mathematician.

"Proving Fermat's Last Theorem again dear?" my wife asked wryly.

"Of course not," I said. "It's been proven, go back to sleep.

Made in the USA
Lexington, KY
26 January 2017